OpenCV 和 Visual Studio
图像识别应用开发

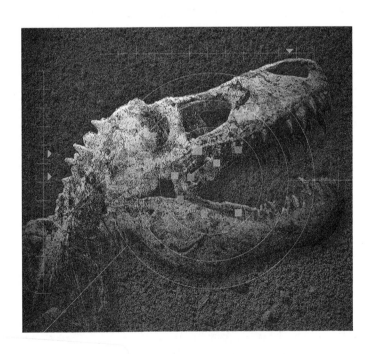

望熙荣 望熙贵 著
李强 改编

人民邮电出版社

北京

图书在版编目（CIP）数据

OpenCV和Visual Studio图像识别应用开发 / 望熙荣，望熙贵著. -- 北京：人民邮电出版社，2017.10
ISBN 978-7-115-46505-4

Ⅰ．①O… Ⅱ．①望… ②望… Ⅲ．①图象处理软件－程序设计②程序语言－程序设计 Ⅳ．①TP391.413 ②TP312

中国版本图书馆CIP数据核字(2017)第203478号

版权声明

原著书名：OpenCV with Microsoft Visual Studio 影像辨识处理
本书中文繁体版本版权由台湾博硕文化股份有限公司（DrMaster Press Co.,Ltd）获作者望熙荣、望熙贵著授权拥有独家出版发行，中文简体版本版权由博硕文化股份有限公司（DrMaster PressCo.,Ltd）获作者同意授权人民邮电出版社独家出版发行。
版权所有，侵权必究。

◆ 著　　望熙荣　望熙贵
　 改　编　李 强
　 责任编辑　陈冀康
　 责任印制　焦志炜

◆ 人民邮电出版社出版发行　北京市丰台区成寿寺路11号
　 邮编　100164　电子邮件　315@ptpress.com.cn
　 网址　http://www.ptpress.com.cn
　 固安县铭成印刷有限公司印刷

◆ 开本：800×1000　1/16
　 印张：18.75　　　　　　　　2017年10月第1版
　 字数：399千字　　　　　　 2024年7月河北第10次印刷

著作权合同登记号　图字：01-2016-6528号

定价：59.00元
读者服务热线：(010)81055410　印装质量热线：(010)81055316
反盗版热线：(010)81055315
广告经营许可证：京东市监广登字20170147号

内容提要

OpenCV 是可以在多平台下运行、并提供了多语言接口的一个库，实现了图像处理和计算机视觉方面的很多通用算法。

本书是介绍 OpenCV 结合 Visual Studio 进行图像识别和处理的编程指南。全书共 11 章，介绍了 OpenCV 和 Visual Studio 的安装设置，以及 Core、HighGUI、ImgProc、Calib3d、Feature2d、Video、Objdetect、ML、Contrib 等模块，涉及文字处理、照片处理、图像识别、OpenGL 整合、硬件设备结合使用等众多方面的功能，最后还给出了综合应用的实例。

本书适合对于图像识别和处理技术感兴趣，并且想要学习 OpenCV 的应用和编程的读者阅读和参考。

序

　　影像处理早期是从扫描的文件中识别出文字（OCR），后来才发展为手写识别、自拍修图等静态图像处理。机器人技术的应用从早期的组装自动化，到中期的生产质量监控，再到近期逐渐走进人群的应用，整个发展过程中都涉及图像处理的技术。因此，图像处理不仅会更加流行，而且会更加普及和接近大众。

　　程序设计在企业中有非常普及的应用，而图像处理则一直是专业人士的领域。如果将两者结合起来，一定能够产生更加广泛的应用领域，大大提高工作效率。然而，直到图像处理的开源软件的兴起，才使得这种结合的可能性越来越大。

　　像 OpenCV 这样开源的软件，虽然使用 C 或 C++来进行开发，但还是要求用户具备基本的编程知识，才能快速上手。对于一般的程序设计人员来说，这已经将图像处理领域的门槛降低了。通过本书的入门级的介绍，以及充分的程序应用实例，读者能够逐渐掌握图像处理的基本编程和应用技能。

　　虽然笔者在编写图书的过程中已经尽力校对，但失误在所难免，恳请广大读者不吝赐教。

作者

目　　录

第1章　系统安装与项目准备 1
- 1.1　认识 OpenCV 2
- 1.2　系统安装 3
  - 1.2.1　安装 OpenCV 3
  - 1.2.2　安装 Visual Studio 2013 6
- 1.3　开始新项目 7
  - 项目属性的设置 9

第2章　Core 模块 16
- 2.1　显示图文件 17
- 2.2　图文件转换 30
- 2.3　图文件混合 32
- 2.4　改变对比与明亮度 39
- 2.5　基本绘图 41
- 2.6　文字处理 49
- 2.7　离散的傅立叶变换 57
- 2.8　使用 XML 与 YAML 进行文件的输出输入 61
- 2.9　与 OpenCV 1 互通 70

第3章　HighGUI 模块 74
- 3.1　滑块功能 75
- 3.2　读取视频文件进行相似性比较 77
- 3.3　产生视频文件 82

第4章　ImgProc 模块 85
- 4.1　图像的平滑化 86
- 4.2　腐蚀与膨胀 90
- 4.3　更多形态处理 93
- 4.4　图像金字塔 96
- 4.5　基本阈值法 99
- 4.6　建立自己的线性滤波器 104
- 4.7　将图像加上边框 107
- 4.8　Sobel 算子 108
- 4.9　拉普拉斯运算 112
- 4.10　Canny 图像边缘检测 113
- 4.11　霍夫线变换 116
- 4.12　霍夫圆变换 121
- 4.13　重映射 124
- 4.14　仿射变换 127
- 4.15　直方图分布平等化 129
- 4.16　直方图分布计算 131
- 4.17　直方图分布比较 134
- 4.18　反向投影 136
- 4.19　模板匹配 149
- 4.20　寻找图的轮廓 153
- 4.21　凸包 159
- 4.22　为轮廓建立许多矩形与圆形 162
- 4.23　为轮廓建立旋转的矩形与椭圆形 164
- 4.24　图像矩 166
- 4.25　点多边形测试 169
- 4.26　线性变换 171

第5章　Calib3d 模块 173
- 5.1　使用棋盘进行相机校准 174
- 5.2　视差 193

第6章　Feature2d 模块 195

6.1 特征描述 ·················· 196
6.2 哈瑞斯角点检测 ············ 198
6.3 使用 FLANN 进行特征
匹配 ···················· 200
6.4 使用 Features2D 和 Homography
识别对象 ················· 202
6.5 Shi-Tomasi 角点检测 ········ 204
6.6 建立自定义的角点检测 ······ 206
6.7 在次像素检测角位置 ········ 209
6.8 特征检测 ················· 211

第 7 章 Video 模块 ··············· 213

7.1 图像拍摄 ················· 214
7.2 生成视频文件 ············· 215
7.3 指定帧 ··················· 218
7.4 移动感知 ················· 219
7.5 计算移动时间 ············· 221
7.6 即时对象追踪 ············· 225
7.7 播放暂停 ················· 229

第 8 章 Objdetect 模块 ··········· 231

级联式类分类 ············· 232

第 9 章 ML 模块 ················· 236

9.1 支持向量机的介绍 ·········· 239
9.2 非线性可分开数据的支持
向量机 ··················· 242

第 10 章 Contrib 模块 ············ 245

探索视网膜效果并用来进
行图像处理 ··············· 246

第 11 章 实际应用 ··············· 251

11.1 图像藏密 ················ 252
11.2 图像采集 ················ 255
11.3 QR Code 检测 ············· 268
11.4 与 OpenGL 整合 ··········· 275

附录 ······························ 281

第 1 章

系统安装与项目准备

图像处理已经流行很多年了。一般大家最早接触的就是 OCR，即由图文件转换成文本文件。但是现在比较热门的图像处理，已经从静态的图像处理转为动态的图像识别，例如，这些技术在机器人或自动驾驶技术等领域的应用都已经较为成熟且很流行。计算机程序设计现在已经十分普遍，但是图像处理却还是较为专业的领域，主要原因是相关知识还不够普及。本书的目的就是希望介绍免费的开源软件，通过清晰的讲解和充分的示例说明，让图像处理技术能够普及，使得其应用更加多元化。

当前，图像识别和处理已经在自动驾驶领域得到了应用，在相关的应用中，汽车不但以图像处理识别赛道，并且加上了 GPS 定位判定，这样即可判断油门加速或减速，最后还能够精准地停回赛道的起跑线。你是否会对这些应用感到惊讶？是不是已经迫不及待要快速掌握图像处理技术呢？

1.1 认识 OpenCV

OpenCV 的全名是 Open Source Computer Vision Library，是 Intel 内部的研究计划，其目的是为了推广 Intel 高端 CPU 的应用。OpenCV 的最初版本是 2000 年在 IEEE 计算机图像与图案识别大会（Computer Vision and Pattern Recognition）中发布的，目前则由非营利的基金组织（OpenCV.org）在负责维护。

OpenCV 的初版主要是以 C 语言作为开发主体，即 OpenCV 1.0。当时这一版本一推出，就造成轰动且极受欢迎，但是最大的问题是，在设计图像处理程序时，程序员必须自行考虑对象的内存管理。如果程序很小，还不会造成问题，但当程序越来越复杂，功能越来越多，再进行对象内存管理就可能会造成麻烦，使得程序的质量变成了大问题，而处理问题所花费的时间比设计程序花的时间还要多。

因此诞生了 OpenCV 2.0 版本。它主要是以 C++开发设计，因为 C++具有类（class），这使得对象的内存管理方便许多。现在，网络上下载的版本代号 2.x.x 就是 OpenCV 2.0 版本。第一个 x 代表小改版，一般是功能的增强或是添加新功能；第二个 x 代表功能的改善或错误的排除。

OpenCV 包含了许多关于计算机图像转换、图像处理以及其它的数学运算处理的功能，它是由许多模块组合而成的，这些模块主要都与图像处理相关。本书将分各章节介绍这些模块。各模块功能如表 1-1 所示。

表 1-1

模 块 组	模 块 功 能
Core	数据类型、数据结构、内存管理
Highgui	读写图形文件、屏幕输入/输出处理、简单的 UI 功能

续表

模 块 组	模 块 功 能
Imgproc	图形滤波（filtering）、几何图形转换、形状分析
Calib3d	照相机校准（Camera calibration）、多视角3D重建
Feature2d	特征提取、描述与对比
Video	视频对象追踪与移动分析
Objdetect	使用级联式（cascade）与方向梯度直方图（histogram-of-gradient）分类器进行对象识别
ML	用于视频处理的统计模式与归类算法
Flann	全名是 Fast Library for Approximate Nearest Neighbors，用于高纬度数据的快速搜索
GPU	以选择的并行算法在 GPU 快速执行
Stitching	视频结合处理（stitching）的方法：弯曲、混合、集束调整（bundle adjustment）
Nonfree	有专利权的算法

NonFree 表示是要付费的，不过到 OpenCV 3.0 这部分模块已经删除。

笔者开始编写本书的时候，OpenCV 3.0 Beta 已在网络上提供下载测试，初步还看不太出来与 2.0 的差异，虽然模块减少了 6 个（contrib、dynamicuda、legacy、nonfree、gpu、ocl），但是增加了 14 个模块（cuda、cudaarithm、cudabgsegm、cudacodec、cudafeatures2d、cudafikters、cudaimgproc、cudalegacy、cudaoptflow、cudastereo、cudawarping、cudev、imgcodecs、shape）。由这些模块的改变看来，3.0 版本应该是架构上有了更改，所以版本才从 2 变成了 3。

如果读者喜欢使用 OpenCV 3.0，可以安装 Ceemple OpenCV，这是最新搭配 Visual Studio 2013 的开发环境，可以节省 OpenCV 安装与项目的设置，详情请参考本书附录。至于新增的 CUDA 相关模块概念，读者可参考 Wrox 出版的《Professional CUDA C Programming》一书，并行处理将是 3.0 的重点，取代了 2.0 的 GPU 模块。

OpenCV 的教程文档（tutorials）说明了如何在 Linux、Mac、iOS 或 Android 上安装，以及如何用 Java、Eclipse、Python 或 Clojure 开发，感兴趣的读者请自行参考。笔者认为 C++ 与 Windows 还是最普遍的，可以在自己最熟悉的环境学会 OpenCV，然后再转换平台或语言就方便、快速了。

1.2 系统安装

1.2.1 安装 OpenCV

读者可以到 OpenCV 的官方网站 http://opencv.org/下载。本书使用的版本是 2.4.10，在

http://opencv.org/downloads.html 下载页面内请单击 OpenCV for Windows，如图 1-1 所示。

图 1-1

下载执行后请输入自解压路径，如图 1-2 所示。
如果要考虑有足够的硬盘空间，也可以改为其它硬盘，接下来的设置则须对应相同的硬盘。
安装完成的文档结构如图 1-3 所示。

图 1-2

图 1-3

安装设置

我们将介绍微软的 C++开发工具 Visual Studio 2013，所以这里的说明是以 Windows 7 为主。请单击 Windows 画面左下角的"开始"，再在"计算机"上单击鼠标右键，然后选取"属性"，如图 1-4 所示。

出现如图 1-5 所示的界面后，单击"高级系统设置"。

图 1-4

图 1-5

画面出现之后再单击"环境变量"按钮，如图 1-6 所示。

环境变量的设置，如图 1-7 所示。

图 1-6

图 1-7

环境变量设置完成的结果如图 1-8 所示。

图 1-8

1.2.2　安装 Visual Studio 2013

Visual Studio Community 2013 是微软首次提供的免费软件，同时此软件提供许多工具供开发各种需求的软件，也支持多语言的开发。本书是以 C++为主，OpenCV 也可以使用 Python 开发。如果读者想使用 Python 开发 OpenCV，可下载 Python Tools for Visual Studio，网址为 https://pytools.codeplex.com/。

下载

读者可到下列网址下载 Visual Studio 2013：https://www.visualstudio.com/downloads/download-visual-studio-vs，如图 1-9 所示。

图 1-9

单击左下角的 DVD9 ISO 图像即可开始下载。如果没有，可以解压缩 ISO 档的软件，可单击上方项目的"立即安装"直接在线进行安装。Visual Studio 2013 安装完成之后，如果要安装中文版本，可单击右侧的语言套件。

笔者建议安装和使用英文版本，如果在开发过程碰到问题时，可以很方便地在国际社区网站上询问，这样就可以知道画面或编译出现的问题描述。提问的网站有 http:// answers.opencv.org/questions/ 与 http://stackoverflow.com/，笔者比较偏好登录 http://stackoverflow.com/。

安装

解压缩完成之后，在解压缩的目录内直接单击"vs_community"即可，如图 1-10 所示。安装完成后，单击"LAUNCH"开始使用，如图 1-11 所示。

图 1-10　　　　　　　　　　　　　　　　图 1-11

这个版本功能强大，但是要求必须注册微软账号，让微软了解软件用户的信息。

1.3　开始新项目

当 Visual Studio 2013 窗口出现之后，单击左侧的"New Project"，如图 1-12 所示。

图 1-12

然后先单击左侧的"Visual C++",再单击右侧的"Win32 Console Application",而项目名称在本范例中是"Display_Image",如图 1-13 所示。

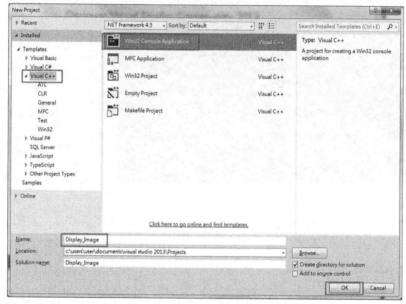

图 1-13

再单击"Next"按钮。因为还有设置要进行,所以不要直接单击"Finish"按钮,如图 1-14 所示。

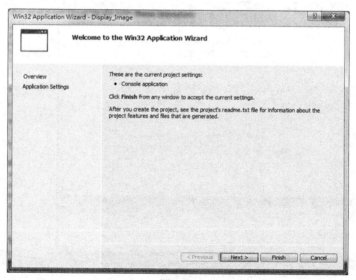

图 1-14

这里要先将已勾选的选项进行清除（清除方式是单击复选框取消勾选），再单击"Empty Project"，此时单击"Finish"按钮，如图 1-15 所示。

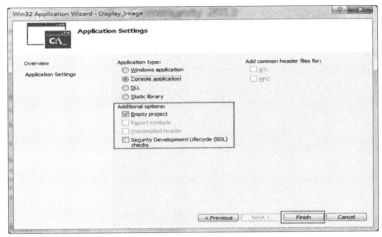

图 1-15

项目属性的设置

项目建立后就要设置项目属性。单击右侧属性窗口内的项目名称，再单击菜单"PROJECT"，然后单击"Properties"，如图 1-16 所示。

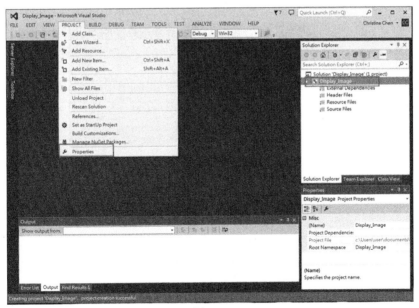

图 1-16

单击"Configuration Properties"左侧的三角形，如图 1-17 所示。

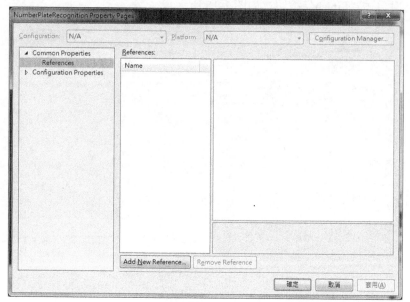

图 1-17

设置"VC++ Directories"

这个部分分为 Include 与 Library，如图 1-18 所示。

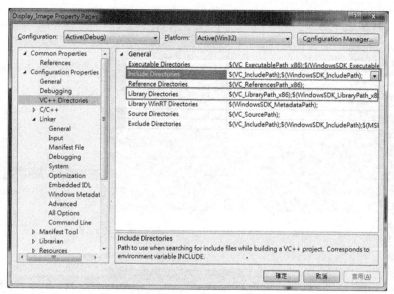

图 1-18

Include 部分

单击"Include Directories"右侧的下拉按钮，会出现如图 1-19 所示的界面。

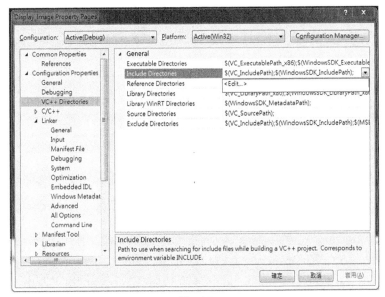

图 1-19

单击下拉列表中的"<Edit...>"，进入如图 1-20 所示的界面。
在窗口上方的空白范围处单击两次，如图 1-21 所示。

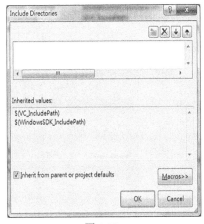

图 1-20　　　　　　　　　　　图 1-21

单击右侧的"..."按钮，然后单击硬盘中安装 OpenCV 的"Include"目录，如图 1-22 所示。
设置完成后如图 1-23 所示。

11

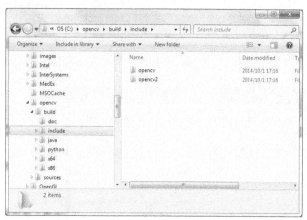

图 1-22

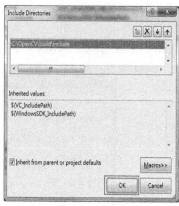

图 1-23

OpenCV 2.0 版本将 1.0 版本所使用的函数名称整个都替换掉了。OpenCV 1.0 是使用 C 语言开发的，所以是传统的 API 概念，而 2.0 是 C++的版本，所以有命名空间与类的概念。不过 OpenCV 1.0 也可以使用本书示范的项目开发方式，只是程序中的 include，在 1.0 时使用 opencv 目录，而 2.0 则是使用 opencv2 的目录。也就是说，使用 OpenCV 1.0 的代码为 #include <opencv/cv.h>，而 OpenCV 2.0 的代码为 #include "opencv2/core/core.hpp"。不过作者测试后发现，在 1.0 程序中使用#include "opencv2/core/core.hpp"也可以正常执行。

Library 部分

画面操作如同 Include，而目录选取内容如图 1-24 所示。个人计算机虽已是 64 位，但还是建议选择 x86，因为这样一来，设置比较简单。

设置完成的结果如图 1-25 所示。

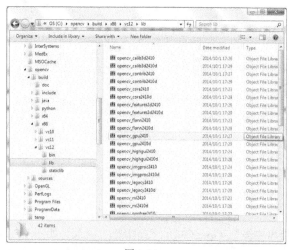

图 1-24

图 1-25

设置连接器（Linker）

操作界面如图 1-26 和图 1-27 所示，这里不再赘述。

这里将本书所使用到的程序库一次全部加入。就本范例而言，只用到前两个程序库（core 与 highgui），其它程序库读者也可待项目使用时再加入。但往往都在碰到连接错误之后才知道要补充程序库。虽然一个模块就有一个程序库文件，但 Visual Studio 发出的编译错误消息对初学者而言不易读懂，从而导致读者不知道要加什么程序库，由此带来困扰，所以我们建议一次性加入所有的程序库。

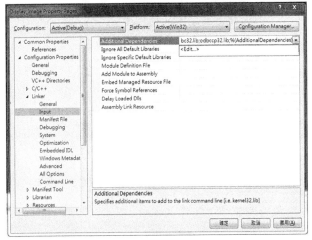

图 1-26

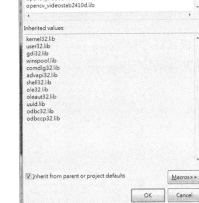

图 1-27

建议读者可以将这些想要加入的程序库内容以 Notepad 程序存储，如图 1-28 所示。下次新项目要使用时，只要用 Notepad 打开文档再复制粘贴即可，可以减少打字时间，也防止录入错误。

到此为止，我们只是完成项目设置，还需要输入程序。此部分请继续阅读第 2 章。

笔者尝试过，无法在 Windows XP 系统上安装 Visual Studio 2013，所以使用 Windows XP 的读者，只能使用 Visual Studio 2010，但所有项目设置可参照 Visual Studio 2013 的说明来进行。

笔者曾在 OpenCV 2.4.8 版时使用 Visual Studio 2010 开发项目，但是当 Windows Update

更新.NET Framework 从 4.0 更新到 4.5 之后，原项目的执行反而会有错误消息，只好将.NET Framework 4.5 删除再重新安装.NET Framework 4.0。所以在使用 Visual Studio 2010 时，要关掉自动更新功能，或是筛选更新内容。

如果卸载 Visual Studio 2013 版本之后又想要再安装，程序会认为没有删除，导致无法重新安装，如图 1-29 所示。

图 1-28

图 1-29

读者只要到 Temp 目录查看安装的记录文件就会得知原因。也就是要将如图 1-30 所示的数据从注册表（registry）中删除。

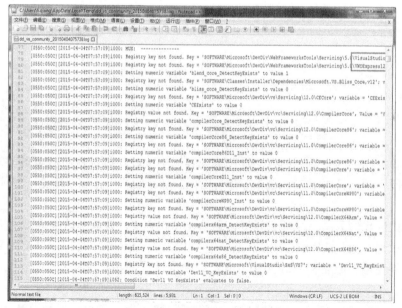

图 1-30

请读者要手动删除 HKEY_LOCAL_MACHINE/SOFTWARE/Wow6432Node/Microsoft/VisualStudio 键下的内容。因为它不会被卸载程序删除，才造成再安装时被误认为原安装仍然存在。只需要删除时使用 Windows 原有的注册表信息即可（单击"开始"→搜索程序及文档→输入 REGEDIT）。

第 2 章

Core 模块

Core 模块是 OpenCV 最基本的模块，因为所有 OpenCV 需要的数据结构与基本的绘图功能都在 Core 模块内，所以在项目中，Core 也是第一个一定要具备的模块。这些基本数据结构与绘图功能，请参考 OpenCV 说明文档的网址 http://docs.opencv.org/。本书将以程序示例来介绍其实际应用，因为任何图书都不可能完全介绍 OpenCV 的所有功能，所以读者要养成经常查看这些说明文档的习惯。只要单击网页内的"core. The Core Functionality"就可以查看 Core 模块。要查看其他模块，方法也是相同的。

2.1 显示图文件

因为 OpenCV 有太多模块，而每个模块又有许多函数，本书就以程序示例的方式来介绍。在程序中使用到的函数，我们才会给出说明，其他未说明的部分，请读者自行查阅 OpenCV 的说明文档。

至此，项目终于设置完成，现在我们开始编写程序。因为我们只是学习操作，所以先剪切并粘贴代码比较简单，其方式如下：

在右侧"Solution Explorer"中的"Source File"中单击鼠标右键，选择"Add"再选择"New Item"来加入代码，如图 2-1 所示。

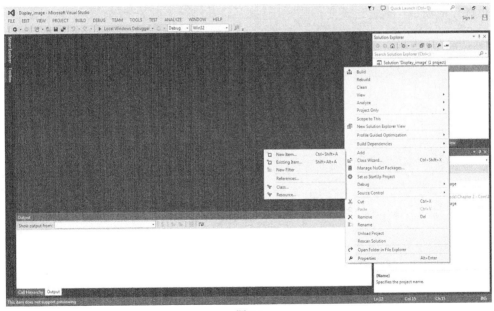

图 2-1

输入程序文件名称，这里使用与项目相同的名称，如图 2-2 所示。

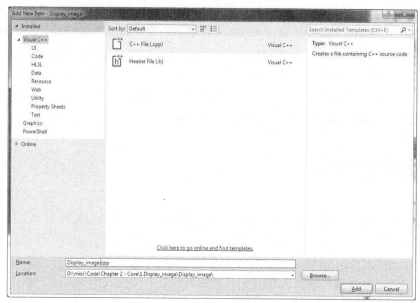

图 2-2

本小节所介绍的代码设置,与其他项目使用的方式都相同,后面所有的项目按照此设置,不再重复说明。

继续先前项目,设置完成有了程序文件之后,就粘贴复制的代码,如图 2-3 所示。

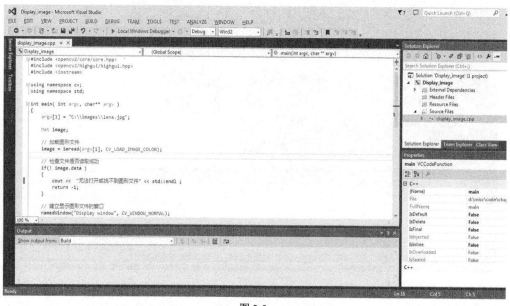

图 2-3

粘贴程序后再单击选单中绿色的三角形按钮，如图 2-4 所示。

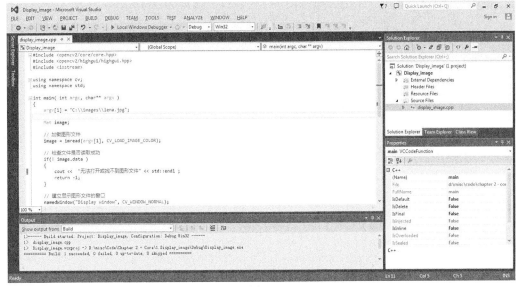

图 2-4

Visual Studio 会询问是否进行编译，单击"Yes"按钮开始执行，如图 2-5 所示。

执行本程序之前，我们应该先将本书所使用的图文件都复制到 C:\images 文件夹下。如果图文件位于不同的目录，请自行修改程序对应的文件夹。

最后程序结果显示，如图 2-6 所示。

图 2-5 　　　　　　　　　　　　　　图 2-6

本项目只有一个程序文件。如果项目有许多 C++程序与 include 文件，请全部复制到这个项目内，如图 2-7 所示。

要注意，就此范例而言，文件应该复制到 Display_Image/Display_Image 文件夹下。

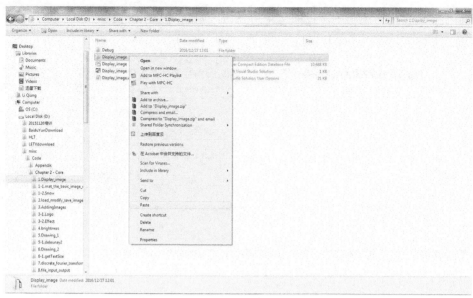

图 2-7

复制完成后再分别将复制的文件加入项目内。加载 Include 文件的方法与加载 C++程序文件的做法一样,只是 Include 文件要加载 Solution Explorer 的 Header File,C++文件要加载 Solution Explorer 的 Source File。

现在以加入 C++程序文件来做说明。在"Solution Explorer"的"Source File"上单击鼠标右键,再依次单击"Add"与"Existing item...",如图 2-8 所示。

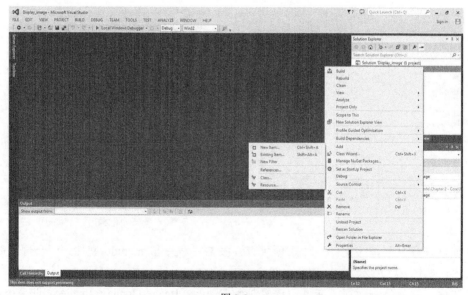

图 2-8

Visual Studio 就会弹出窗口让你选择要加入的内容，如图 2-9 所示。

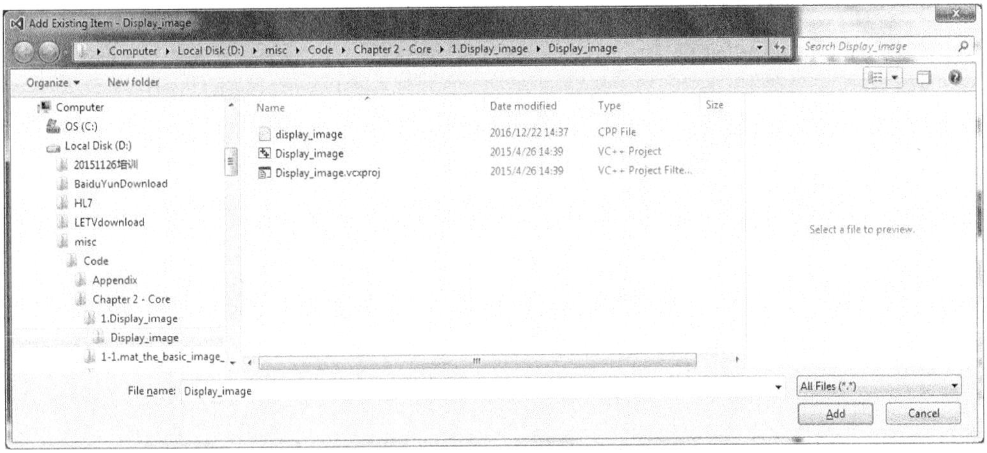

图 2-9

这只是 OpenCV 的"Hello World"（第一个测试程序的意思），所以直接写入要处理的文件（本书范例全都是直接写入的状态）。读者也可以自行与 MFC 或 C++/CLI 整合来处理 UI 问题，这部分不是本书探讨的范畴。

原始文件都是以 argv 作为输入文件，可以改变文件名，或在项目属性的"Debugging"选项中设置，如图 2-10 所示。笔者认为直接写入程序并且印在书中，读者在阅读时则能看得清程序的结构。

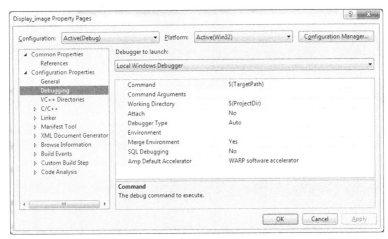

图 2-10

注意

因为 Windows 7 在复制文件路径时会自动变成反斜杠，所以这里继续沿用，如图 2-11 所示。为了避免 C++ 程序编译错误，所以改成双反斜杠，如果使用正斜杠就不需要再使用双正斜杠了，如图 2-12 所示。

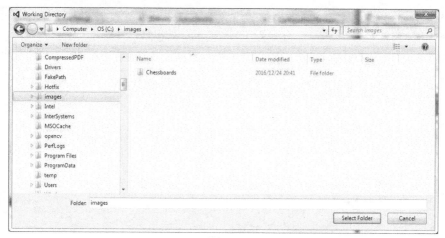

图 2-11

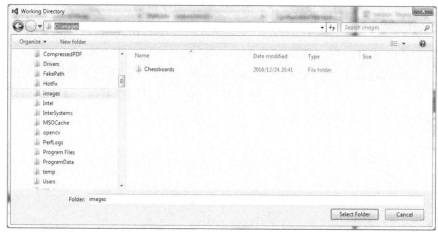

图 2-12

程序内容如下：

```
#include <opencv2/core/core.hpp>
#include <opencv2/highgui/highgui.hpp>
#include <iostream>

using namespace cv;
using namespace std;

int main( int argc, char** argv )
{
    argv[1] = "C:\\images\\lena.jpg";

    Mat image;
```

```
// 载入图文件
image = imread(argv[1], CV_LOAD_IMAGE_COLOR);

// 检查读取文件是否成功
if(! image.data )
{
    cout <<  "无法打开或找不到图文件" << std::endl ;
     return -1;
}

// 建立显示图文件窗口
namedWindow("Display window", CV_WINDOW_NORMAL);

// CV_WINDOW_FREERATIO 与 CV_WINDOW_KEEPRATIO
// CV_GUI_NORMAL 与 CV_GUI_EXPANDED

// 在窗口内显示图文件
imshow( "Display window", image );

// 窗口等待按键
waitKey(0);

return 0;
}
```

程序说明

本书先假设读者已对 C++有所接触认识，代码内容属于 C++而不是 OpenCV 的范畴，在此不再赘述。

1. Mat image

Mat 是 OpenCV 新定义的数据类型，类似传统的数据类型 int、float 或 String。Mat 这种数据类型表示图像，而图像都是二维数组。所以 OpenCV 就定义了处理图像的矩阵类别（Matrix），取英文的前 3 个字母 Mat，就如同 int 取 integer 的前 3 个字母一样。

OpenCV 常用的数据结构有以下几种。

- Point：代表二维的点，用于图像的坐标点。

    Point pt;
    pt.x = 10;
    pt.y = 8;

或是：

    Point pt = Point(10, 8)

- Scalar：代表 4 元素的向量（4-element vector），一般用于 RGB 颜色值。

    Scalar(a, b, c)，第 4 个参数如果用不到可以省略。
    a 代表蓝色值、b 代表绿色值、c 代表红色值，也就是 Scalar(B,G,R)。

MAT 类别一般运算的说明见表 2-1。

表 2-1

语 法	说 明
double m[2][2] = {{1.0, 2.0}, {3.0, 4.0}}; Mat M(2, 2, CV_32F, m);	由多维数组（array）产生 2×2 矩阵
Mat M(100, 100, CV_32FC2, Scalar(1, 3));	声明 100×100 的两个通道的矩阵，第一个通道值是 1，第二个通道值是 3
M.create(300, 300, CV_8UC(15));	产生 300×300 的 15 个通道的矩阵，之前变量 M 内的值将被删除
Mat ima(240, 320, CV_8U, Scalar(100))	产生 240×320，单一通道（灰阶）的矩阵
int sizes[3] = {7, 8, 9}; Mat M(3, sizes, CV_8U, Scalar::all(0));	产生三维矩阵，各维的维度值（dimension）分别是 7、8 和 9，矩阵的初始值是 0
Mat M = Mat::eye(7, 7, CV_32F);	产生 32 位浮点（float）的 7×7 的矩阵
Mat M = Mat::zeros(7, 7, CV_64F);	产生 64 位浮点（float）的 7×7 的矩阵，矩阵初始值是 0
M.at<double>(i, j)	取得矩阵 M 在第 i 行、第 j 列的值，该值的数据类型是 double。 注意：行列的初始值都是 0
M.row(1)	取得矩阵 M 在第 1 行的值
M.col(3)	取得矩阵 M 在第 3 列的值
M.rowRange(1, 4)	取得矩阵 M 在第 1 行到第 4 行的值
M.colRange(1, 4)	取得矩阵 M 在第 1 列到第 4 列的值
M.rowRange(2, 5).colRange(1, 3)	取得矩阵 M 在第 2 行到第 5 行及第 1 列到第 3 列的值
M.diag();	取得矩阵 M 的对角值
Mat M2 = M1.clone();	将 M1 复制给 M2
Mat M2; M1.copyTo(M2);	将 M1 复制给 M2
Mat M1 = Mat::zeros(9, 3, CV_32FC3); Mat M2 = M2 = M1.reshape(0, 3);	声明 3 行的 M2 矩阵，其通道与 M1 相同(reshape 的第一个参数 0)，3 行的数组是 reshape 的第二个参数 3
Mat M2 = M1.t();	M2 是 M1 的转置（transpose）
Mat M2 = M1.inv();	M2 是 M1 的逆矩阵（inverse）
Mat M3 = M1 * M2;	M3 是 M1 乘 M2（矩阵相乘）
Scalar s; Mat M2 = M1 + s;	M2 是 M1 矩阵加上 s 颜色值（scalar）
Mat image; image.size().height image.size().width	Image 矩阵的高和宽

Mat 是数据类型，就会被当成参数在函数之间传递。但是因为图文件可能非常庞大，为了节省内存与性能，OpenCV 都是以传址的方式处理的，也就是共用一份内存。现在用下面的代码来说明这一点。

```
Mat A, C;
A = imread(argv[1], CV_LOAD_IMAGE_COLOR);
C = A;
Mat fun(A);
```

上面的程序声明了两个图像数据类型 A 和 C，A 的内容是读入的图文件，然后 A 又传递 C（C = A），A 再以参数传递给 fun 函数。如果 fun 函数对 A 做任何处理，可能改变了 A，fun 函数执行完 C 也会被改变，即使 C 不在 fun 函数的作用域之内。

不同数据类型名称如果共用内存的话，何时才会真的从内存消失？OpenCV 是使用 C++ 的引用计数（reference couting），所以我们在设计程序时不用考虑做内存管理，否则就要执行对象释放（Release）的操作，这就是 OpenCV 2.0 与 OpenCV 1.0 的差别。

2. imread (const string& filename, int flags)：**读取图文件**

   filename：图文件名称

   flags：读取的方式。本例是读取彩色的内容。

   OpenCV 可以读取的图文件种类如下。
   - BMP：Windows 位图文件。
   - PBM、PGM、PPM：可移植图文件格式。
   - SR、RAS：Sun 的图文件格式。
   - JPEG、JPG、JPE：JPEG 图文件格式。
   - TIFF、TIF：TIFF 图文件格式。
   - PNG：可移植网络图文件格式。

   读取的方式有下列 3 种选项：可使用如下例程名称，或用数值表示。
   - CV_LOAD_IMAGE_UNCHANGED：这是以图文件的原始文件形态（彩色或黑白）读取，使用数值是小于零的数值。
   - CV_LOAD_GRAYSCALE：这是以黑白方式读取，使用数值是等于零的数值。
   - CV_LOAD_COLOR：这是以彩色方式读取，使用数值是大于零的数值。

3. namedWindow (const string& winname, int flags)：**建立要显示图文件的窗口**

   winname：窗口名称。这也表示可以同时建立多个窗口，再以名称区别

   flags 有下列几种选择。
   - WINDOW_AUTOSIZE：这是窗口大小不可改变，会按照图文件的大小自动调整；如果使用 CV_WINDOW_AUTOSIZE 也可以达到同样效果，这是 OpenCV 1.0 的用法
   - WINDOW_NORMAL：这是窗口大小可以改变，但必须是使用 Qt 开发；有关使用 C++

与 Qt 的说明，请参考 http://qt-project.org/wiki/OpenCV_with_Qt。
- WINDOW_OPENGL：设置支持 OpenGL 时，就要使用此参数。
  不过笔者现阶段测试失败了，但是在 C:\OpenCV\sources\modules\highgui\src\window_w32.cpp 程序中查找 OPENGL 部分，系统的确是有支持 OpenGL 的功能，可能 OpenCV 在构建程序库时有参数设置遗漏了。
- CV_WINDOW_NORMAL：如果重新建立对 OpenGL 的支持，想要再次使用 WINDOW_AUTOSIZE 就会出错，此时需要使用此参数。这可以使窗口大小随意改变，放大不会导致图像失真，缩小也不会影响原图内容比例。
  这个参数在没有支持 OpenGL 时也可以使用，该选项特别适用于窗口上有滑动杆，而滑动杆大于窗口宽度时，或图太小想放大窗口看清效果。

4. imshow(const string& winname, InputArray mat)：显示窗口

  winname：窗口名称，会显示在窗口的左上方，方便窗口之间的分辨或比较。
  mat：图文件所声明的矩阵参数。
  本程序示范显示图像窗口，当程序结束时窗口就会自动关闭。如果程序还有其他部分在使用中，但又想关闭窗口时，则可以使用 destroyWindow() 来关闭窗口，或使用 destroyAllWindows() 来关闭所有窗口。

5. waitKey()：等待按键

  这是等待用户按下任意键继续执行，否则程序会立刻结束，并且无法看到程序执行结果。其参数是以毫秒（ms）为单位的等待时间。如果参数值为零，则表示持续等待。
  注意，这个等待功能只能用于 imshow 所显示的窗口，若没有任何使用 imshow 显示的窗口，那 waitKey() 将没有任何作用。此时可改为用传统的 getchar() 实现程序执行的暂停功能。
  如果将鼠标光标在图像的窗口上移动，光标会自动变成十字形，这表示 OpenCV 在窗口处理上还可以处理鼠标。第 11 章的实际案例应用的图像采集就是鼠标光标的示例。

图像数据结构实例

学习 OpenCV，除了函数之外，就是数据结构最重要了，所以我们再以程序范例来对此加以说明。不过一般图像矩阵是无法用 << 来显示内容的，本程序只是以读者看得懂的方式来说明。

```
#include "opencv2/core/core.hpp"
#include <iostream>

using namespace std;
using namespace cv;

int main(int, char**)
{
```

```cpp
// 用构造函数建立数据
Mat M(2, 2, CV_8UC3, Scalar(0, 0, 255));
cout << "M = " << endl << " " << M << endl;

// 用 create 函数建立数据
M.create(2, 2, CV_8UC(2));
cout << "M = " << endl << " " << M << endl;

// 建立多维矩阵
int sz[3] = { 2, 2, 2 };
Mat L(3, sz, CV_8UC(1), Scalar::all(0));
// 无法用 << 运算符输出

// 用MATLAB风格的眼建立数据
Mat E = Mat::eye(3, 3, CV_64F);
cout << "E = " << endl << " " << E << endl;

// 数据都是1
Mat O = Mat::ones(2, 2, CV_32F);
cout << "O = " << endl << " " << O << endl;

// 数据都是0
Mat Z = Mat::zeros(2, 2, CV_8UC1);
cout << "Z = " << endl << " " << Z << endl;

// 建立 3x3 双精确度矩阵，值由<<输入
Mat C = (Mat_<double>(3, 3) << 0, -1, 0, -1, 5, -1, 0, -1, 0);
cout << "C = " << endl << " " << C << endl << endl;

// 复制第一行数据
Mat RowClone = C.row(1).clone();
cout << "RowClone = " << endl << " " << RowClone << endl;

// 以随机数值填入矩阵内
Mat R = Mat(3, 2, CV_8UC3);
randu(R, Scalar::all(0), Scalar::all(255));

// 展示各种输出格式选项
cout << "R (default) = " << endl << R << endl;
cout << "R (python)  = " << endl << format(R, "python") << endl;
cout << "R (numpy)   = " << endl << format(R, "numpy") << endl;
cout << "R (csv)     = " << endl << format(R, "csv") << endl;
cout << "R (c)       = " << endl << format(R, "C") << endl;

// 图像中二维的点
Point2f P(5, 1);
cout << "Point (2D) = " << P << endl << endl;

// 图像中三维的点
Point3f P3f(2, 6, 7);
cout << "Point (3D) = " << P3f << endl << endl;

vector<float> v;
v.push_back((float)CV_PI);
v.push_back(2);
```

```
        v.push_back(3.01f);

    cout << "浮点向量矩阵 = " << Mat(v) << endl << endl;

    vector<Point2f> vPoints(5);
    for (size_t i = 0; i < vPoints.size(); ++i)
        vPoints[i] = Point2f((float)(i * 5), (float)(i % 7));

    cout << "二维图点向量 = " << vPoints << endl;

    getchar();
    return 0;
}
```

执行结果如图 2-13 所示。

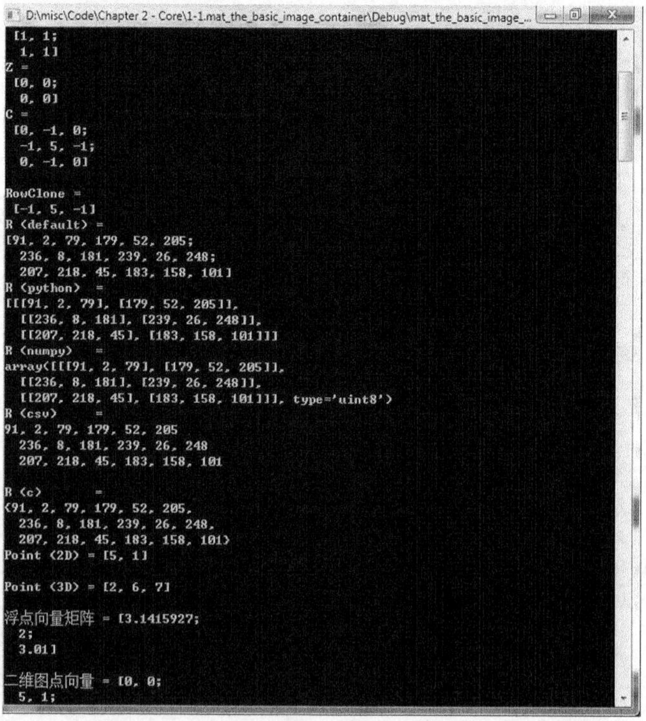

图 2-13

下雪特效

下雪特效的代码如下：

```
#include <opencv2/core/core.hpp>
#include <opencv2/highgui/highgui.hpp>
#include <iostream>

using namespace cv;
```

```cpp
using namespace std;

int main(int argc, char** argv)
{
    argv[1] = "C:\\images\\lake.jpg";

    Mat image;

    // 载入图文件
    image = imread(argv[1], CV_LOAD_IMAGE_COLOR);

    // 检查读取文件是否成功
    if (!image.data)
    {
      cout << "无法打开或找不到图文件" << std::endl;
      return -1;
    }

    // 建立显示图文件窗口
    namedWindow("原图", CV_WINDOW_NORMAL);
    namedWindow("下雪图", CV_WINDOW_NORMAL);

    imshow("原图", image);

    // 雪点数
    int i = 600;

    for (int k = 0; k < i; k++) {
        // rand() is the MFC random number generator
        // try qrand() with Qt
        int i = rand() % image.cols;
        int j = rand() % image.rows;
        if (image.channels() == 1) { // gray-level image
           image.at<uchar>(j, i) = 255;
           if (i < (int)image.cols)
              image.at<uchar>(j + 1, i) = 255;
           if (j < (int)image.rows)
              image.at<uchar>(j, i + 1) = 255;
           if (i < (int)image.cols && j < (int)image.rows)
              image.at<uchar>(j + 1, i + 1) = 255;
        }
        else if (image.channels() == 3) { // color image
           image.at<cv::Vec3b>(j, i)[0] = 255;
           image.at<cv::Vec3b>(j, i)[1] = 255;
           image.at<cv::Vec3b>(j, i)[2] = 255;
           if (i < (int)image.cols - 1)
           {
               image.at<cv::Vec3b>(j, i + 1)[0] = 255;
               image.at<cv::Vec3b>(j, i + 1)[1] = 255;
               image.at<cv::Vec3b>(j, i + 1)[2] = 255;
           }
           if (j < (int)image.rows - 1)
           {
               image.at<cv::Vec3b>(j + 1, i)[0] = 255;
               image.at<cv::Vec3b>(j + 1, i)[1] = 255;
```

```
                image.at<cv::Vec3b>(j + 1, i)[2] = 255;
            }
            if (j < (int)image.rows - 1 && i < (int)image.cols - 1)
            {
                image.at<cv::Vec3b>(j + 1, i + 1)[0] = 255;
                image.at<cv::Vec3b>(j + 1, i + 1)[1] = 255;
                image.at<cv::Vec3b>(j + 1, i + 1)[2] = 255;
            }
        }
    }

    // 在窗口内显示图文件
    imshow("下雪图", image);

    // 窗口等待按键
    waitKey(0);

    return 0;
}
```

程序执行的结果对比, 如图 2-14 所示。

（a）原图

（b）飘雪图

图 2-14

本程序是示范图像矩阵内每一个点的处理。

2.2 图文件转换

图文件转换的代码如下：

```
#include <opencv\cv.h>
#include <opencv\highgui.h>

using namespace cv;

int main(int argc, char* argv)
```

```cpp
{
    // 图文件
    char* imageName = "C:\\images\\lena.jpg";

    // 读取图文件
    Mat image = imread(imageName, 1);

    Mat gray_image;
    // 图文件从 BGR 转成灰度
    cvtColor(image, gray_image, CV_BGR2GRAY);

    // 存储转换后的图文件
    imwrite("C:\\images\\process\\Gray_lena.jpg", gray_image);

    // 显示图文件窗口大小的控制
    namedWindow(imageName, CV_WINDOW_AUTOSIZE);
    namedWindow("Gray image", CV_WINDOW_AUTOSIZE);

    // 显示原先图文件
    imshow(imageName, image);

    // 显示灰度图文件
    imshow("C:\\images\\process\\Gray image", gray_image);

    waitKey(0);

    return 0;
}
```

程序说明

1. cvtColor (InputArray src, OutputArray dst, int code, int dstCn=0)：图像颜色空间 (color space) 转换

    （1）src：输入图像。
    （2）dst：输出图像。
    （3）code：颜色空间转换种类。
    - RGB 与 CIE 转换：CV_BGR2XYZ、CV_RGB2XYZ、CV_XYZ2BGR、CV_XYZ2RGB。
    - RGB 与 YCrCB JPEG 转换：CV_BGR2YCrCb、CV_RGB2YCrCb、CV_YCrCb2BGR、CV_YCrCb2RGB。
    - RGB 与 HSV 转换：CV_BGR2HSV、CV_RGB2HSV、CV_HSV2BGR、CV_HSV2RGB。
    - RGB 与 HLS 转换：CV_BGR2HLS、CV_RGB2HLS、CV_HLS2BGR、CV_HLS2RGB。
    - RGB 与 CIE L*a*b*转换：CV_BGR2Lab、CV_RGB2Lab、CV_Lab2BGR、CV_Lab2RGB。
    - RGB 与 CIE L*u*v 转换：CV_BGR2Luv、CV_RGB2Luv、CV_Luv2BGR、CV_Luv2RGB。
    - Bayer 转换成 RGB：CV_BayerBG2BGR、CV_BayerGB2BGR、CV_BayerRG2BGR、CV_BayerGR2BGR、CV_BayerBG2RGB、CV_BayerGB2RGB、CV_BayerRG2RGB、CV_BayerGR2RGB。

（4）dstCn：输出图像通道数，如果值为 0，输出图像通道数由输入图像 src 与颜色空间 code 自动取得。

2. imwrite(const string& filename, InputArray img)：**存储图像**

（1）filename：要存储的文件名。

（2）img：要存储的图像。

程序执行结果，图 2-15（a）为原图彩色，图 2-15（b）为转换后的灰度图。

（a）原图彩色　　　　　　　　　　（b）灰度图

图 2-15

2.3　图文件混合

图文件混合的代码如下：

```
#include "opencv2/highgui/highgui.hpp"
#include <iostream>

using namespace cv;

int main(void)
{
    double alpha, beta, input;

    Mat src1, src2, dst;

    /// 让用户输入 alpha 值
    std::cout << " 简易线性混合(Linear Blender) " << std::endl;
    std::cout << "------------------------" << std::endl;
    std::cout << "* 输入 0 到 1 的 alpha 值: ";
    std::cin >> input;

    // 确认 alpha 值数入的正确在于 0 与 1 之间
```

```
    if (alpha >= 0 && alpha <= 1)
    {
     alpha = input;
    }

    /// 读取两个大小与类型相同的图文件
    src1 = imread("C:\\images\\LinuxLogo.jpg");
    src2 = imread("C:\\images\\WindowsLogo.jpg");

    if (!src1.data)
    { std::cout << "读取 src1 错误" << std::endl; return -1; }

    if (!src2.data)
    { std::cout << "读取 src2 错误" << std::endl; return -1; }

    namedWindow("Linear Blend", 1);

    beta = (1.0 - alpha);
    addWeighted(src1, alpha, src2, beta, 0.0, dst);

    imshow("Linear Blend", dst);

    waitKey(0);
    return 0;
}
```

程序说明

addWeighted(InputArray src1, double alpha, InputArray src2, double beta, double gamma, OutputArray dst, int dtype=-1)：以权重将两图合并

（1）src1：要相加的第一个图文件。

（2）alpha：第一个图文件的权重。

（3）src2：要相加的第二个图文件，因为只是单纯的与第一个图文件相加，所以大小与通道数要与第一个图文件相同。

（4）beta：第二个图文件的权重。

（5）gamma：两图相加后要再增加的值。

（6）dst：两图相加结果的图文件。

（7）dtype：相加结果图的景深（depth）；此参数可有可无。

本程序是使用 addWeighted 函数实现下列公式的图像处理：

$$g(x)=(1-\alpha)f_0(x)+\alpha f_1(x);$$

式中，$f_0(x)$ 是第一个图；$f_1(x)$ 是第二个图，因为 addWeighted 在 OpenCV 内部就是 dst=α·src1+β·src2+γ，只是公式中的 γ 为 0。

输入分别为 0、0.5 和 1，对应的执行结果如图 2-16 所示。

图 2-16

读者如果使用过 Photoshop 的图层，应该对此功能不陌生。

注意

虽然 alpha 与 beta 是两张图的权重，但其实可以是各小于 1 的值，而不用相加等于 1。

商标特效

商标特效的代码如下：

```
#include <opencv2/core/core.hpp>
#include <opencv2/highgui/highgui.hpp>
#include "opencv2/imgproc/imgproc.hpp"
#include <vector>

using namespace cv;

int main()
{
    // 载入图文件
    Mat image1 = imread("C:\\images\\lena.jpg");
    Mat logo =
        imread("C:\\opencv\\build\\doc\\opencv-logo-white.png");

    // 编译器要求使用前要给初始值
    Mat image = image1, opencvlogo;

    // 缩小原图成 Size(col, row)
    resize(logo, opencvlogo, Size(80, 64));

    namedWindow("Image 1", CV_WINDOW_AUTOSIZE);

    // 定义图有兴趣的区域(Region Of Interest, ROI)
    Mat imageROI;

    // 指定商标在原图的位置，Rect(x, y, width(col), height(row))
    imageROI = image(Rect(420, 420, 80, 64));

    imshow("Image 1", opencvlogo);

    // 加入商标
    addWeighted(imageROI, 1.0, opencvlogo, 0.3, 0., imageROI);

    // 显示结果
```

```
    namedWindow("with logo");
    imshow("with logo", image);

    waitKey();

    return 0;
}
```

程序说明

1. resize(InputArray src, OutputArray dst, Size dsize, double fx=0, double fy=0, int interpolation=INTER_LINEAR)：改变图像大小

（1）src：输入图像。
（2）dst：输出图像。
（3）dsize：输出图像的图像大小，如果等于 0 就用下列公式计算。
dsize = Size(round(fx*src.cols), round(fy*src.rows)，但是 fx 与 fy 都不可等于 0。
（4）fx：水平轴的缩放因子，如果等于 0 就用下列公式计算：
(double)dsize.width/src.cols。
（5）fy：垂直轴的缩放因子，如果等于 0 就用下列公式计算：
(double)dsize.height/src.rows。
（6）interpolation：插值方式。
- INTER_NEAREST：最靠近周围插值法（nearest-neighbor）。
- INTER_LINEAR：双线性插值法（bilinear）。
- INTER_AREA：用像素关系区再取样插值法（resampling）。
- INTER_CUBIC：在 4×4 像素附近用双立方插值法（bicubic）。
- INTER_LANCZOS4：在 8×8 像素附近用 Lanczos 插值法。

2. size Mat::size() const：返回矩阵大小

size(cols, rows)也返回矩阵大小，同时指定矩阵的行列数。

执行结果如图 2-17 所示。

图 2-17

addWeighted(src1, 0.7, src2, 0.9, 0.0, dst)又可以改写成如下的代码，根据读者喜好自行决定。

```
src1 = src1 * 0.7;
src2 = src2 * 0.9;
add(src1, src2, dst);
```

也可简单写成 dst = src1 * 0.7 + src2 * 0.9，只是 Add 函数还有其他的参数可以使用。

雨天特效

雨天特效的代码如下：

```cpp
#include <opencv2/core/core.hpp>
#include <opencv2/highgui/highgui.hpp>
#include "opencv2/imgproc/imgproc.hpp"
#include <vector>

using namespace cv;

int main()
{
    Mat image1, image2, image3;

    image1 = imread("C:\\images\\lake.jpg");
    if (!image1.data)
        return 0;

    image2= imread("C:\\images\\fur.jpg");
    if (!image2.data)
        return 0;

    // 以 image2 图像大小调整 image1 图像大小
    resize(image2, image3, image1.size());

    //显示原图
    namedWindow("Image 1");
    imshow("Image 1",image1);

    namedWindow("Image 3");
    imshow("Image 3",image3);

    // 雨天特效图
    Mat result;

    image3 = image3 * 0.3;
    image1 = image1 * 0.9;
    add(image1, image3, result);

    namedWindow("result");
    imshow("result",result);

    waitKey();

    return 0;
}
```

程序说明

add(InputArray src1, InputArray src2, OutputArray dst, InputArray mask=noArray(), int dtype=-1)：计算两图像或颜色值相加

（1）src1：第一个图像或颜色值。

（2）src2：第二个图像或颜色值，与第一个图像同大小。

(3) dst：输出图像。
(4) mask：掩码（mask），可有可无。
(5) dtype：输出图像的景深，可有可无。

湖的原图和特效图如图 2-18 和图 2-19 所示。

图 2-18

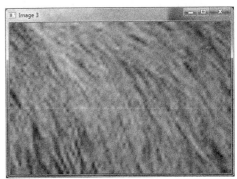

图 2-19

雨天特效结果如图 2-20 所示。

特效图其实是从熊身上采集来的图，如图 2-21 所示。

图 2-20

图 2-21

本节程序是介绍图文件像素（pixel）的存储与处理。在说明程序之前，我们先介绍计算机系统中色彩的概念。像素是由颜色空间（color space）或通道（channel）与数据类型（data type）来描述的。对于最简单的黑白图像，我们调整颜色空间与数据类型就可以产生各种灰度的灰色图。

OpenCV 像素的数据类型是以下列方式表达的：CV_ABCD。

- A：每个像素多少位。
- B：是否有正负号。
- C：类型前置码。

- D：通道数目。

例如，CV_8UC3 表示如下。

A：每个像素为 8 位。

B：没有正负号。

C：因为 8 位没有正负号，所以使用 Char 来表示像素。

D：每个像素有 3 个通道。

RGB 三原色使用 CV_8UC3 表示如下。

(255, 0, 0)：红色。

(0, 255, 0)：绿色。

(0, 0, 255)：蓝色。

(0, 0, 0)：黑色。

(255, 255, 255)：白色。

所谓彩色就有很多方式的应用，利用各种的组合让颜色更多样化。一般色彩系统分为以下 4 种方式。

- RGB：它是最普遍也是与肉眼最接近的，由红（R）、绿（G）、蓝（B）三色组成，屏幕显示都是用此系统，但注意通道的顺序是 RGB。
- HSV 与 HLS：系统将颜色分为色调（hue）、饱和度（saturation）和明亮度（luminance），是最自然说明色彩的方式。
- YCrCb：主要用于 JPEG 图像。
- CIE L*a*b*：这是概念上单一的色彩空间，用来计算色距（distance of color）最为方便。

扫描图文件快慢的比较

一般图都是 R、G、B 三原色。如果颜色太多就会造成图文件非常大，所以为了降低图文件大小都会采用褪色（color reduction）。褪色的目的是加快图文件的扫描进行图像处理，还要加快处理的速度。读者可参考 C:\OpenCV\sources\samples\cpp\tutorial_code\core\how_to_scan_images 的范例来了解该用法快慢的差异。

OpenCV 的遮罩处理（mask operation）

这是根据遮罩值重算图像中的每个像素。例如图像处理方式的公式如下：

$$I(i,j) = 5 \times I(i,j) - [I(i-1,j) + I(i+1,j) + I(i,j-1) + I(i,j+1)]$$

而遮罩的方式则是：

$$I(i,j) \times M,\ \text{其中}\ M = \begin{array}{c|ccc} i\backslash j & -1 & 0 & +1 \\ \hline -1 & 0 & -1 & 0 \\ 0 & -1 & 5 & -1 \\ +1 & 0 & -1 & 0 \end{array}$$

OpenCV 提供 filter2D 函数来进行遮罩处理。

读者可以参考 C:\OpenCV\sources\samples\cpp\tutorial_code\core\mat_mask_operations 的范例程序来了解如何进行遮罩处理。

2.4 改变对比与明亮度

改变对比与明亮度的代码如下：

```cpp
#include <opencv/cv.h>
#include <opencv/highgui.h>
#include <iostream>
using namespace cv;

double alpha;  // 对比控制
int beta;      // 明亮度控制

int main(int argc, char** argv)
{
    Mat image = imread("C:\\images\\lena.jpg");

    Mat new_image = Mat::zeros(image.size(), image.type());

    /// Initialize values (Basic Linear Transform)
    std::cout << " 基本线性转换" << std::endl;
    std::cout << "--------------------------" << std::endl;
    std::cout << "* 输入 alpha 值 [1.0-3.0]: "; std::cin >> alpha;
    std::cout << "* 输入 beta 值 [0-100]: "; std::cin >> beta;

    /// 转换公式 new_image(i,j) = alpha*image(i,j) + beta
    for (int y = 0; y < image.rows; y++)
    {
        for (int x = 0; x < image.cols; x++)
        {
            for (int c = 0; c < 3; c++)
            {
                // 针对像素的每个通道做转换
                new_image.at<Vec3b>(y, x)[c] =
                    saturate_cast<uchar>(alpha*(image.at<Vec3b>(y, x)[c]) + beta);
            }
        }
    }

    namedWindow("Original Image", 1);
    namedWindow("New Image", 1);

    imshow("Original Image", image);
    imshow("New Image", new_image);

    waitKey();

    return 0;
}
```

程序说明

图像转换可以使用两种方式：
- 点（point）的处理。
- 区块（Neighborhood area-based）处理。

明亮与对比的调整就属于点的处理，而点的处理最普遍就是加乘处理。

$$g(x)=\alpha f(x)+\beta$$

```
Mat new_image = Mat::zeros( image.size(), image.type() );
```

为了不影响原图，先建立一个大小与原图相同的矩阵，内容先补 0。

```
new_image.at<Vec3b>(y,x)[c] =
    saturate_cast<uchar>( alpha*( image.at<Vec3b>(y,x)[c] ) + beta );
```

图像处理使用 3 个 for 循环处理图像的每个点，第一个是行、第二个是列、第三个是通道。因为原图是彩色的，所以处理 3 个通道（R、G、B）需要 3 个循环。由于是对图像的每一个点进行处理，所以时间就会久一点。

在图像点的指定是使用 image.at<Vec3b>(y,x)[c]，y 是列、x 是行，c 就是 R、G、B（0、1、2），R = 0，以此类推。

这 3 层循环程序也可以简化成为如下单一指令。

```
image.convertTo(new_image,-1, alpha, beta);
```

执行结果 alpha = 2.2，beta = 50，如图 2-22 所示。

图 2-22

2.5 基本绘图

基本绘图的代码如下:

```cpp
#include <opencv2/core/core.hpp>
#include <opencv2/imgproc/imgproc.hpp>
#include <opencv2/highgui/highgui.hpp>

#define w 400

using namespace cv;

/// 函数声明
void MyEllipse(Mat img, double angle);
void MyFilledCircle(Mat img, Point center);
void MyPolygon(Mat img);
void MyLine(Mat img, Point start, Point end);

int main(void){

  /// 窗口名称
  char atom_window[] = "Drawing 1: Atom";
  char rook_window[] = "Drawing 2: Rook";

  /// 建立空的黑图
  Mat atom_image = Mat::zeros(w, w, CV_8UC3);
  Mat rook_image = Mat::zeros(w, w, CV_8UC3);

  /// 1. 简单绘图
  /// -----------------------

  /// 1.a. 绘制椭圆
  MyEllipse(atom_image, 90);
  MyEllipse(atom_image, 0);
  MyEllipse(atom_image, 45);
  MyEllipse(atom_image, -45);

  /// 1.b. 绘制圆
  MyFilledCircle(atom_image, Point(w / 2, w / 2));

  /// 2. 绘制城堡
  /// ------------------

  /// 2.a. 绘制凸面多边形
  MyPolygon(rook_image);

  /// 2.b. 绘制长方形
  rectangle(rook_image,
          Point(0, 7 * w / 8),
          Point(w, w),
          Scalar(0, 255, 255),
          -1,
          8);
```

```cpp
    /// 2.c. 绘制直线
    MyLine(rook_image, Point(0, 15 * w / 16), Point(w, 15 * w / 16));
    MyLine(rook_image, Point(w / 4, 7 * w / 8), Point(w / 4, w));
    MyLine(rook_image, Point(w / 2, 7 * w / 8), Point(w / 2, w));
    MyLine(rook_image, Point(3 * w / 4, 7 * w / 8), Point(3 * w / 4, w));

    // 显示窗口
    imshow(atom_window, atom_image);

    //将窗口移动到指定位置
    moveWindow(atom_window, 0, 200);

    imshow(rook_window, rook_image);
    moveWindow(rook_window, w, 200);

    waitKey(0);

    return(0);
}

// 绘制椭圆
void MyEllipse(Mat img, double angle)
{
    int thickness = 2;
    int lineType = 8;

    ellipse(img,
        Point(w / 2, w / 2),
        Size(w / 4, w / 16),
        angle,
        0,
        360,
        Scalar(255, 0, 0),
        thickness,
        lineType);
}

// 绘制圆
void MyFilledCircle(Mat img, Point center)
{
    int thickness = -1;
    int lineType = 8;

    circle(img,
        center,
        w / 32,
        Scalar(0, 0, 255),
        thickness,
        lineType);
}

// 绘制凸面多边形
void MyPolygon(Mat img)
{
    int lineType = 8;

    /** 先建立点 */
    Point rook_points[1][20];
```

```cpp
    rook_points[0][0] = Point(w / 4, 7 * w / 8);
    rook_points[0][1] = Point(3 * w / 4, 7 * w / 8);
    rook_points[0][2] = Point(3 * w / 4, 13 * w / 16);
    rook_points[0][3] = Point(11 * w / 16, 13 * w / 16);
    rook_points[0][4] = Point(19 * w / 32, 3 * w / 8);
    rook_points[0][5] = Point(3 * w / 4, 3 * w / 8);
    rook_points[0][6] = Point(3 * w / 4, w / 8);
    rook_points[0][7] = Point(26 * w / 40, w / 8);
    rook_points[0][8] = Point(26 * w / 40, w / 4);
    rook_points[0][9] = Point(22 * w / 40, w / 4);
    rook_points[0][10] = Point(22 * w / 40, w / 8);
    rook_points[0][11] = Point(18 * w / 40, w / 8);
    rook_points[0][12] = Point(18 * w / 40, w / 4);
    rook_points[0][13] = Point(14 * w / 40, w / 4);
    rook_points[0][14] = Point(14 * w / 40, w / 8);
    rook_points[0][15] = Point(w / 4, w / 8);
    rook_points[0][16] = Point(w / 4, 3 * w / 8);
    rook_points[0][17] = Point(13 * w / 32, 3 * w / 8);
    rook_points[0][18] = Point(5 * w / 16, 13 * w / 16);
    rook_points[0][19] = Point(w / 4, 13 * w / 16);

    const Point* ppt[1] = { rook_points[0] };
    int npt[] = { 20 };

    fillPoly(img,
        ppt,
        npt,
        1,
        Scalar(255, 255, 255),
        lineType);
}

// 绘制直线
void MyLine(Mat img, Point start, Point end)
{
    int thickness = 2;
    int lineType = 8;
    line(img,
        start,
        end,
        Scalar(0, 0, 0),
        thickness,
        lineType);
}
```

程序说明

1. rectangle(Mat& img, Point pt1, Point pt2, const Scalar& color, int thickness=1, int lineType=8, int shift=0)：绘制矩形

或是：

rectangle(Mat& img, Rect rec, const Scalar& color, int thickness=1, int lineType=8, int shift=0)

（1）img：要绘制矩形所在的图像。

(2) pt1：矩形的顶点。

(3) pt2：矩形 pt1 对角的顶点。

(4) color：矩形的颜色。

(5) thickness：矩形线的厚度。

(6) lineType：矩形线的格式。

(7) shift：点坐标内的部分位数。

(8) rec：要绘制的矩形内容。

2. ellipse(Mat& img, Point center, Size axes, double angle, double startAngle, double endAngle, const Scalar& color, int thickness=1, int lineType=8, int shift=0)：绘制椭圆

或是：

ellipse(Mat& img, const RotatedRect& box, const Scalar& color, int thickness=1, int lineType=8)

(1) img：要绘制椭圆的图像。

(2) center：椭圆圆心。

(3) axes：椭圆主轴（main axes）的一半。

(4) angle：椭圆旋转角度度数。

(5) startAngle：椭圆弧度起始角度。

(6) endAngle：椭圆弧度结束角度。

(7) color：椭圆的颜色。

(8) thickness：椭圆弧线的厚度。

(9) lineType：绘制椭圆线的形态。

- 8：8 的连接线（8-connected line）。
- 4：4 的连接线（4-connected line）。
- CV_AA：非别名线（antialiased line）。

(10) shift：椭圆心与轴值（values of axes）坐标的部分位数。

(11) box：椭圆外围矩形。

3. circle(Mat& img, Point center, int radius, const Scalar& color, int thickness=1, int lineType=8, int shift=0)：绘制圆

(1) img：要绘制圆的图像。

(2) center：圆的圆心。

(3) radius：圆的半径。

(4) color：圆的颜色。

(5) thickness：圆线的厚度。

(6) lineType：绘制圆线的形态。

（7）shift：圆心与轴值（values of axes）坐标的部分位数。

4. fillPoly(Mat& img, const Point** pts, const int* npts, int ncontours, const Scalar& color, int lineType=8, int shift=0, Point offset=Point())：将多边形围绕区域填满

（1）img：要填满区域所在的图像。
（2）pts：多边形数组，数组内容是多边形点的数组。
（3）npts：多边形顶点数数组。
（4）ncontours：要填满区域的轮廓数。
（5）color：多边形的颜色。
（6）lineType：多边形边的形态。
（7）shift：顶点坐标（vertex coordinate）的部分位数。
（8）offset：轮廓的位移点。

5. line(Mat& img, Point pt1, Point pt2, const Scalar& color, int thickness=1, int lineType=8, int shift=0)：绘制直线

（1）img：要绘制直线的图像。
（2）pt1：直线的起点。
（3）pt2：直线的终点。
（4）color：直线的颜色。
（5）thickness：直线的厚度。
（6）lineType：直线的形态。
（7）shift：点坐标（point coordinate）内的部分位数（fractional bit）。

本程序是示范 Point 与 Scalar（色数）两个结构的应用。Point 就是二维图像的点，Scalar 在 OpenCV 普遍用于彩色像素的表达，Scalar(B, G, R)，也就是以颜色数表达图像的每个像素。

执行结果如图 2-23 所示。

 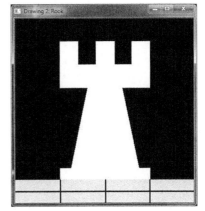

图 2-23

绘制架构图

绘制架构图的代码如下:

```cpp
#include <opencv2/imgproc/imgproc.hpp>
#include <opencv2/highgui/highgui.hpp>
#include <iostream>

using namespace cv;
using namespace std;

static void draw_subdiv_point( Mat& img, Point2f fp, Scalar color )
{
    circle( img, fp, 3, color, CV_FILLED, 8, 0 );
}

static void draw_subdiv( Mat& img, Subdiv2D& subdiv,
        Scalar delaunay_color )
{
#if 1
    vector<Vec6f> triangleList;
    subdiv.getTriangleList(triangleList);
    vector<Point> pt(3);

    for( size_t i = 0; i < triangleList.size(); i++ )
    {
        Vec6f t = triangleList[i];
        pt[0] = Point(cvRound(t[0]), cvRound(t[1]));
        pt[1] = Point(cvRound(t[2]), cvRound(t[3]));
        pt[2] = Point(cvRound(t[4]), cvRound(t[5]));
        line(img, pt[0], pt[1], delaunay_color, 1, CV_AA, 0);
        line(img, pt[1], pt[2], delaunay_color, 1, CV_AA, 0);
        line(img, pt[2], pt[0], delaunay_color, 1, CV_AA, 0);
    }
#else
    vector<Vec4f> edgeList;
    subdiv.getEdgeList(edgeList);
    for( size_t i = 0; i < edgeList.size(); i++ )
    {
        Vec4f e = edgeList[i];
        Point pt0 = Point(cvRound(e[0]), cvRound(e[1]));
        Point pt1 = Point(cvRound(e[2]), cvRound(e[3]));
        line(img, pt0, pt1, delaunay_color, 1, CV_AA, 0);
    }
#endif
}

static void locate_point( Mat& img, Subdiv2D& subdiv, Point2f fp,
                    Scalar active_color )
{
    int e0=0, vertex=0;

    subdiv.locate(fp, e0, vertex);

    if( e0 > 0 )
    {
```

```cpp
        int e = e0;
        do
        {
            Point2f org, dst;
            if( subdiv.edgeOrg(e, &org) > 0 &&
                subdiv.edgeDst(e, &dst) > 0 )
                line( img, org, dst, active_color, 3, CV_AA, 0 );

            e = subdiv.getEdge(e, Subdiv2D::NEXT_AROUND_LEFT);
        }
        while( e != e0 );
    }

    draw_subdiv_point( img, fp, active_color );
}

static void paint_voronoi( Mat& img, Subdiv2D& subdiv )
{
    vector<vector<Point2f> > facets;
    vector<Point2f> centers;
    subdiv.getVoronoiFacetList(vector<int>(), facets, centers);

    vector<Point> ifacet;
    vector<vector<Point> > ifacets(1);

    for( size_t i = 0; i < facets.size(); i++ )
    {
        ifacet.resize(facets[i].size());
        for( size_t j = 0; j < facets[i].size(); j++ )
            ifacet[j] = facets[i][j];

        Scalar color;
        color[0] = rand() & 255;
        color[1] = rand() & 255;
        color[2] = rand() & 255;
        fillConvexPoly(img, ifacet, color, 8, 0);

        ifacets[0] = ifacet;
        polylines(img, ifacets, true, Scalar(), 1, CV_AA, 0);
        circle(img, centers[i], 3, Scalar(), -1, CV_AA, 0);
    }
}


int main( int, char** )
{
    Scalar active_facet_color(0, 0, 255),
           delaunay_color(255,255,255);

    Rect rect(0, 0, 600, 600);

    Subdiv2D subdiv(rect);
    Mat img(rect.size(), CV_8UC3);
```

```
        img = Scalar::all(0);
        string win = "Delaunay Demo";
        imshow(win, img);

        for( int i = 0; i < 200; i++ )
        {
            Point2f fp( (float)(rand()%(rect.width-10)+5),
                        (float)(rand()%(rect.height-10)+5));

            locate_point( img, subdiv, fp, active_facet_color );
            imshow( win, img );

            if( waitKey( 100 ) >= 0 )
                break;

            subdiv.insert(fp);

            img = Scalar::all(0);
            draw_subdiv( img, subdiv, delaunay_color );
            imshow( win, img );

            if( waitKey( 100 ) >= 0 )
                break;
        }

        img = Scalar::all(0);
        paint_voronoi( img, subdiv );
        imshow( win, img );

        waitKey(0);

        return 0;
    }
```

程序说明

polylines(Mat& img, const Point** pts, const int* npts, int ncontours, bool isClosed, const Scalar& color, int thickness=1, int lineType=8, int shift=0)：绘制多边形曲线

或是：

polylines(InputOutputArray img, InputArrayOfArrays pts, bool isClosed, const Scalar& color, int thickness=1, int lineType=8, int shift=0)

（1）img：输入图像。
（2）pts：多边形曲线点的数组。
（3）npts：多边形曲线顶点数的数组。
（4）ncontours：曲线数。
（5）isClosed：表示是否为封闭的多边形。
（6）color：多边形曲线的颜色。

（7）thickness：多边形曲线的厚度。
（8）lineType：多边形曲线的形态。
（9）shift：顶点坐标部分位数（fractional bits）。

执行结果如图 2-24 所示。

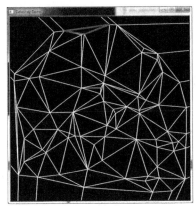 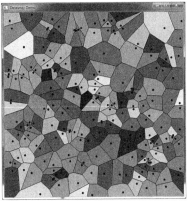

图 2-24

2.6 文字处理

文字处理的代码如下：

```
#include <opencv2/core/core.hpp>
#include <opencv2/imgproc/imgproc.hpp>
#include <opencv2/highgui/highgui.hpp>
#include <iostream>
#include <stdio.h>

using namespace cv;

/// 声明全局变量
const int NUMBER = 100;
const int DELAY = 5;

const int window_width = 900;
const int window_height = 600;
int x_1 = -window_width/2;
int x_2 = window_width*3/2;
int y_1 = -window_width/2;
int y_2 = window_width*3/2;

/// 声明函数
static Scalar randomColor( RNG& rng );
int Drawing_Random_Lines( Mat image, char* window_name, RNG rng );
int Drawing_Random_Rectangles( Mat image, char* window_name, RNG rng );
int Drawing_Random_Ellipses( Mat image, char* window_name, RNG rng );
int Drawing_Random_Polylines( Mat image, char* window_name, RNG rng );
```

```cpp
int Drawing_Random_Filled_Polygons( Mat image, char* window_name, RNG rng );
int Drawing_Random_Circles( Mat image, char* window_name, RNG rng );
int Displaying_Random_Text( Mat image, char* window_name, RNG rng );
int Displaying_Big_End( Mat image, char* window_name, RNG rng );

int main( void )
{
  int c;

  /// 窗口名称
  char window_name[] = "Drawing_2 Tutorial";

  /// 建立随机对象（RNG），并设置起始值
  RNG rng( 0xFFFFFFFF );

  /// 声明内容为 0 的矩阵
  Mat image = Mat::zeros( window_height, window_width, CV_8UC3 );

  /// 显示窗口
  imshow( window_name, image );
  waitKey( DELAY );

  /// 绘制直线
  c = Drawing_Random_Lines(image, window_name, rng);
  if( c != 0 ) return 0;

  /// 绘制长方形
  c = Drawing_Random_Rectangles(image, window_name, rng);
  if( c != 0 ) return 0;

  /// 绘制椭圆
  c = Drawing_Random_Ellipses( image, window_name, rng );
  if( c != 0 ) return 0;

  /// 绘制连接线（polylines）
  c = Drawing_Random_Polylines( image, window_name, rng );
  if( c != 0 ) return 0;

  /// 绘制多边形
  c = Drawing_Random_Filled_Polygons( image, window_name, rng );
  if( c != 0 ) return 0;

  /// 绘制圆
  c = Drawing_Random_Circles( image, window_name, rng );
  if( c != 0 ) return 0;

  /// 绘制文字
  c = Displaying_Random_Text( image, window_name, rng );
  if( c != 0 ) return 0;

  /// 结局
  c = Displaying_Big_End( image, window_name, rng );
  if( c != 0 ) return 0;

  waitKey(0);
```

```cpp
    return 0;
}

/// 函数定义

// 随机产生颜色
static Scalar randomColor( RNG& rng )
{
  int icolor = (unsigned) rng;
  return Scalar( icolor&255, (icolor>>8)&255, (icolor>>16)&255 );
}


// 绘制直线
int Drawing_Random_Lines( Mat image, char* window_name, RNG rng )
{
  Point pt1, pt2;

  for( int i = 0; i < NUMBER; i++ )
  {
    pt1.x = rng.uniform( x_1, x_2 );
    pt1.y = rng.uniform( y_1, y_2 );
    pt2.x = rng.uniform( x_1, x_2 );
    pt2.y = rng.uniform( y_1, y_2 );

    line( image, pt1, pt2, randomColor(rng), rng.uniform(1, 10), 8 );
    imshow( window_name, image );
    if( waitKey( DELAY ) >= 0 )
      { return -1; }
  }

  return 0;
}

// 绘制长方形
int Drawing_Random_Rectangles( Mat image, char* window_name, RNG rng )
{
  Point pt1, pt2;
  int lineType = 8;
  int thickness = rng.uniform( -3, 10 );

  for( int i = 0; i < NUMBER; i++ )
  {
    pt1.x = rng.uniform( x_1, x_2 );
    pt1.y = rng.uniform( y_1, y_2 );
    pt2.x = rng.uniform( x_1, x_2 );
    pt2.y = rng.uniform( y_1, y_2 );

    rectangle( image, pt1, pt2, randomColor(rng), MAX( thickness, -1 ), lineType );

    imshow( window_name, image );
    if( waitKey( DELAY ) >= 0 )
      { return -1; }
  }
```

```cpp
    return 0;
}

// 绘制椭圆
int Drawing_Random_Ellipses( Mat image, char* window_name, RNG rng )
{
    int lineType = 8;

    for ( int i = 0; i < NUMBER; i++ )
    {
        Point center;
        center.x = rng.uniform(x_1, x_2);
        center.y = rng.uniform(y_1, y_2);

        Size axes;
        axes.width = rng.uniform(0, 200);
        axes.height = rng.uniform(0, 200);

        double angle = rng.uniform(0, 180);

        ellipse( image, center, axes, angle, angle - 100, angle + 200,
                 randomColor(rng), rng.uniform(-1,9), lineType );

        imshow( window_name, image );

        if( waitKey(DELAY) >= 0 )
          { return -1; }
    }

    return 0;
}

// 绘制多边形(polylines)
int Drawing_Random_Polylines( Mat image, char* window_name, RNG rng )
{
    int lineType = 8;

    for( int i = 0; i< NUMBER; i++ )
    {
        Point pt[2][3];
        pt[0][0].x = rng.uniform(x_1, x_2);
        pt[0][0].y = rng.uniform(y_1, y_2);
        pt[0][1].x = rng.uniform(x_1, x_2);
        pt[0][1].y = rng.uniform(y_1, y_2);
        pt[0][2].x = rng.uniform(x_1, x_2);
        pt[0][2].y = rng.uniform(y_1, y_2);
        pt[1][0].x = rng.uniform(x_1, x_2);
        pt[1][0].y = rng.uniform(y_1, y_2);
        pt[1][1].x = rng.uniform(x_1, x_2);
        pt[1][1].y = rng.uniform(y_1, y_2);
        pt[1][2].x = rng.uniform(x_1, x_2);
        pt[1][2].y = rng.uniform(y_1, y_2);

        const Point* ppt[2] = {pt[0], pt[1]};
        int npt[] = {3, 3};
```

```cpp
    polylines(image, ppt, npt, 2, true, randomColor(rng), rng.uniform(1,10), lineType);

    imshow( window_name, image );
    if( waitKey(DELAY) >= 0 )
       { return -1; }
  }
  return 0;
}

// 绘制多边形
int Drawing_Random_Filled_Polygons( Mat image, char* window_name, RNG rng )
{
  int lineType = 8;

  for ( int i = 0; i < NUMBER; i++ )
  {
    Point pt[2][3];
    pt[0][0].x = rng.uniform(x_1, x_2);
    pt[0][0].y = rng.uniform(y_1, y_2);
    pt[0][1].x = rng.uniform(x_1, x_2);
    pt[0][1].y = rng.uniform(y_1, y_2);
    pt[0][2].x = rng.uniform(x_1, x_2);
    pt[0][2].y = rng.uniform(y_1, y_2);
    pt[1][0].x = rng.uniform(x_1, x_2);
    pt[1][0].y = rng.uniform(y_1, y_2);
    pt[1][1].x = rng.uniform(x_1, x_2);
    pt[1][1].y = rng.uniform(y_1, y_2);
    pt[1][2].x = rng.uniform(x_1, x_2);
    pt[1][2].y = rng.uniform(y_1, y_2);

    const Point* ppt[2] = {pt[0], pt[1]};
    int npt[] = {3, 3};

    fillPoly( image, ppt, npt, 2, randomColor(rng), lineType );

    imshow( window_name, image );
    if( waitKey(DELAY) >= 0 )
        { return -1; }
  }
  return 0;
}

// 绘制圆
int Drawing_Random_Circles( Mat image, char* window_name, RNG rng )
{
  int lineType = 8;

  for (int i = 0; i < NUMBER; i++)
  {
    Point center;
    center.x = rng.uniform(x_1, x_2);
    center.y = rng.uniform(y_1, y_2);

    circle( image, center, rng.uniform(0, 300), randomColor(rng),
```

```cpp
                rng.uniform(-1, 9), lineType );
    imshow( window_name, image );
    if( waitKey(DELAY) >= 0 )
      { return -1; }
  }

  return 0;
}

// 绘制文字
int Displaying_Random_Text( Mat image, char* window_name, RNG rng )
{
  int lineType = 8;

  for ( int i = 1; i < NUMBER; i++ )
  {
    Point org;
    org.x = rng.uniform(x_1, x_2);
    org.y = rng.uniform(y_1, y_2);

    putText( image, "Testing text rendering", org,
             rng.uniform(0,8), rng.uniform(0,100)*0.05+0.1,
             randomColor(rng), rng.uniform(1, 10), lineType);

    imshow( window_name, image );
    if( waitKey(DELAY) >= 0 )
      { return -1; }
  }

  return 0;
}

// 测试是否可以显示中文
int Displaying_Big_End( Mat image, char* window_name, RNG )
{
  char end_text[] = "我爱 OpenCV!";
  // 取得文字大小
  Size textsize = getTextSize(end_text, FONT_HERSHEY_COMPLEX, 3, 5, 0);
  Point org((window_width - textsize.width)/2,
    (window_height - textsize.height)/2);

  int lineType = 8;

    Mat image2;

  for( int i = 0; i < 255; i += 2 )
  {
    // 原图扣除所有颜色、饱和度（saturate）的处理
    image2 = image - Scalar::all(i);

    putText(image2, end_text, org,
            FONT_HERSHEY_COMPLEX + FONT_ITALIC, 3,
            Scalar(i, i, 255), 4, lineType);
```

```
        imshow( window_name, image2 );

        if( waitKey(DELAY) >= 0 )
        { return -1; }
    }

    return 0;
}
```

执行结果如图 2-25 所示。

图 2-25

程序说明

1. RNG(uint64 state)：产生随机数

　　state：起始随机数序（random sequence）的 64 位值。

　　本程序是示范用随机数（RNG）产生绘图的内容（颜色与位置），同时用同一个窗口显示内容并用 waitKey(DELAY)来作为显示间的区隔，所以可以看到图不会在窗口内清除，直到最后用 for loop 来清除底图。

2. Size getTextSize (const string& text, int fontFace, double fontScale, int thickness, int* baseLine)：计算文字符串的长与宽

　　（1）text：输入的文字符串。
　　（2）fontFace：字体；此字体将用于 putText()函数。
　　（3）fontScale：字体大小；与 putText 相同。
　　（4）thickness：字体笔画粗细；与 putText 相同。

（5）baseLine：文字底线的 y 坐标。

此函数用于算出文字用参数的字体与笔画粗细所涵盖的方形区块。

3. putText (Mat& img, const string& text, Point org, int fontFace, double fontScale, Scalar color, int thickness, int lineType, bool bottomLeftOrigin)：在图像中绘制文字字符串

（1）img：要显示文字的图。
（2）text：要显示的文字。
（3）org：文字在图中的左下角。
（4）fontFace：字体；可用的字体如下：

```
FONT_HERSHEY_SIMPLEX, FONT_HERSHEY_PLAIN,
FONT_HERSHEY_DUPLEX, FONT_HERSHEY_COMPLEX,
FONT_HERSHEY_TRIPLEX, FONT_HERSHEY_COMPLEX_SMALL,
FONT_HERSHEY_SCRIPT_SIMPLEX,
FONT_HERSHEY_SCRIPT_COMPLEX,
```

字体可以与 FONT_ITALIC 组合

（5）fontScale：字体的大小。
（6）color：文字的颜色。
（7）thickness：文字笔划的粗细度。
（8）lineType：文字笔划的形态，此参数与画线的函数相同。
（9）bottomLeftOrigin：此参数可选，若有此参数（true 值），字符串图的起点在左下角，否则为右上角。

getTextSize 与 putText 的应用说明请参考下列程序将更清楚。

笔者在程序中尝试用各种字体来显示中文，但似乎不能处理双字节的文字。这部分 OpenCV 可能尚无改进的计划，除非更改源代码为 Unicode 再重建 OpenCV 的程序库。但是现在要重建 OpenCV 似乎不可能了，因为许多 OpenCV 内置利用到的软件已被并购。在 Windows 环境重建可以参考说明文档：http://docs.opencv.org/doc/tutorials/introduction/windows_install/windows_install.html#installation-by-making-your-own-libraries-from-the-source-files。

文字部分代码示范

文字部分代码如下：

```
#include <opencv2/core/core.hpp>
#include <opencv2/highgui/highgui.hpp>
#include <iostream>
#include <stdio.h>

using namespace cv;

int main(void)
{
    string text = "Funny text inside the box";
```

```cpp
    int fontFace = FONT_HERSHEY_SCRIPT_SIMPLEX;
    double fontScale = 2;
    int thickness = 3;

    Mat img(300, 900, CV_8UC3, Scalar::all(0));

    int baseline = 0;
    Size textSize = getTextSize(text, fontFace,
     fontScale, thickness, &baseline);
    baseline += thickness;

    // 计算文字中间点
    Point textOrg((img.cols - textSize.width) / 2,
        (img.rows + textSize.height) / 2);

    // 在图像中绘制方盒子
    rectangle(img, textOrg + Point(0, baseline),
        textOrg + Point(textSize.width, -textSize.height),
        Scalar(0, 0, 255));

    // 在图像中绘制文字底线
    line(img, textOrg + Point(0, thickness),
        textOrg + Point(textSize.width, thickness),
        Scalar(0, 0, 255));

    // 在图像中放置文字
    putText(img, text, textOrg, fontFace, fontScale,
            Scalar::all(255), thickness, 8);

    imshow("getText", img);

    waitKey();
}
```

执行结果如图 2-26 所示。

图 2-26

2.7 离散的傅立叶变换

离散的傅立叶变换代码如下:

```
#include "opencv2/core/core.hpp"
#include "opencv2/imgproc/imgproc.hpp"
```

```cpp
#include "opencv2/highgui/highgui.hpp"

#include <iostream>

using namespace cv;
using namespace std;

int main(int argc, char ** argv)
{
    const char* filename = "C:\\images\\lena.jpg";
    Mat I = imread(filename, CV_LOAD_IMAGE_GRAYSCALE);
    if (I.empty())
        return -1;

    Mat padded;

    // 扩张图像到理想的大小
    int m = getOptimalDFTSize(I.rows);
    int n = getOptimalDFTSize(I.cols);

    // 在图像边框补 0 值
    copyMakeBorder(I, padded, 0, m - I.rows, 0,
        n - I.cols, BORDER_CONSTANT, Scalar::all(0));

    Mat planes[] = { Mat_<float>(padded),
        Mat::zeros(padded.size(), CV_32F) };

    Mat complexI;

    // 使用 0 增加到另一个扩增面
    merge(planes, 2, complexI);

    // 将结果大小与原图吻合
    dft(complexI, complexI);

    // 计算扩增值并更换成对数的刻度
    // => log(1 + sqrt(Re(DFT(I))^2 + Im(DFT(I))^2))
    // planes[0] = Re(DFT(I), planes[1] = Im(DFT(I))
    split(complexI, planes);

    // planes[0] = magnitude
    magnitude(planes[0], planes[1], planes[0]);
    Mat magI = planes[0];

    // 更换成对数的刻度
    magI += Scalar::all(1);
    log(magI, magI);

    // 如果有奇数的行或列就修剪光谱 (spectrum)
    magI = magI(Rect(0, 0, magI.cols & -2, magI.rows & -2));

    // 安排傅立叶图的四象限区块 (quadrants)
    // 以便原点位在图的中间
    int cx = magI.cols / 2;
    int cy = magI.rows / 2;
```

```
    // 左上 - 每区块都产生图块
    Mat q0(magI, Rect(0, 0, cx, cy));

    // 右上
    Mat q1(magI, Rect(cx, 0, cx, cy));

    // 左下
    Mat q2(magI, Rect(0, cy, cx, cy));

    // 右下
    Mat q3(magI, Rect(cx, cy, cx, cy));

    Mat tmp;

    // 四象限区块左上与右下对调
    q0.copyTo(tmp);
    q3.copyTo(q0);
    tmp.copyTo(q3);

    // 四象限区块右上与左下对调
    q1.copyTo(tmp);
    q2.copyTo(q1);
    tmp.copyTo(q2);

    // 以浮点值（0与1之间）转换图像到可见的图
    normalize(magI, magI, 0, 1, CV_MINMAX);

    // Show the result
    imshow("Input Image", I);
    imshow("spectrum magnitude", magI);

    waitKey(0);

    return 0;
}
```

程序说明

1. int getOptimalDFTSize(int vecsize)：由输入的向量大小取得最佳的DFT(离散的傅立叶变换)大小

    vecsize：输入的向量大小。

2. copyMakeBorder(InputArray src, OutputArray dst, int top, int bottom, int left, int right, int borderType, const Scalar& value=Scalar())：在图像外围增加边框

    （1）src：原图像。
    （2）dst：结果的图像；大小是原图加上下左右扩展的像素数。
    （3）top：向上扩展的像素数。
    （4）bottom：向下扩展的像素数。
    （5）left：向左扩展的像素数。

（6）right：向右扩展像素数。

（7）borderType：边框类型，具体如下。

```
BORDER_CONSTANT、BORDER_DEFAULT、BORDER_ISOLATED
BORDER_REFLECT、BORDER_REFLECT101、BORDER_REFLECT_101
BORDER_REPLICATE、BORDER_TRANSPARENT
BORDER_WRAP
```

（8）value：如果边框类形是 BORDER_CONSTANT，此参数代表边框值。

3. dft(InputArray src, OutputArray dst, int flags=0, int nonzeroRows=0)：执行向前或反向的离散傅立叶变换

（1）src：输入图像。

（2）dst：结果图像。

（3）flags：转换标志，可以是下列组合。

- DFT_INVERSE：执行反向转换。
- DFT_SCALE：缩放结果。
- DFT_ROWS：对输入图像的每一列执行向前或反向转换。
- DFT_COMPLEX_OUTPUT：执行向前转换。
- DFT_REAL_OUTPUT：执行一维或二维复数数组（complex array）反向转换。

4. magnitude(InputArray x, InputArray y, OutputArray magnitude)：计算二维向量的级数 (magnitude)

（1）x：向量 x 坐标的浮点数组。

（2）y：向量 y 坐标的浮点数组。

（3）magnitude：输出数组，大小和类型与第一参数相同。

5. log(InputArray src, OutputArray dst)：对数运算

（1）src：输入图像。

（2）dst：输出图像。

6. normalize(InputArray src, OutputArray dst, double alpha=1, double beta=0, int norm_type=NORM_L2, int dtype=-1, InputArray mask=noArray()）：数组值距（value range）或基准（norm）的正常化处理

或是：

normalize(const SparseMat& src, SparseMat& dst, double alpha, int normType)

（1）src：输入图像。

（2）dst：输出图像。

（3）alpha：基准正常化（norm normalization）的基准值，或是范围正常化（range

normalization）的较低范围边界值（lower range boundary）。

（4）beta：范围正常化的较高范围边界值（upper range boundary），但不用于基准正常化。

（5）norm_type：正常化类形。

（6）dtype：若是负值，输出与输入有相同的类型；若是非负值，输出与输入有相同的通道与深度。

（7）mask：可有可无的遮罩值。

执行结果如图 2-27 所示。

图 2-27

傅立叶变换是将图像拆解成正弦与余弦元素，也就是将图由空间性（spatial）转换成频率性（frequency），这个概念是任何函数可由无数的正弦与余弦函数加总。傅立叶变换图的数学模式是：

$$F(k,l) = \sum_{i=0}^{N-1} \sum_{j=0}^{N-1} f(i,j) e^{-i2\pi\left(\frac{ki}{N}+\frac{lj}{N}\right)}$$

$$e^{ix} = \cos x + i \sin x$$

其中，$f(i,j)$ 就是图的空间值，$F(k,l)$ 就是图的频率值。

2.8 使用 XML 与 YAML 进行文件的输出输入

本范例是通过 XML 或 YAML 文件来访问 OpenCV 的数据结构，XML 规范请参考官方说明文档（http://www.w3c.org/XML）；YAML 规范请参考官方说明文档（http://www.yaml.org）。

通过 XML 或 YAML 文件来访问 OpenCV 的数据结构的代码如下：

```cpp
#include <opencv2/core/core.hpp>
#include <iostream>
#include <string>

using namespace cv;
using namespace std;

// 定义数据结构
class MyData
{
public:
    MyData() : A(0), X(0), id()
    {}

    // 开放以避免隐性转换
    explicit MyData(int) : A(97), X(CV_PI), id("mydata1234")
    {}

    // 此类的输出
    void write(FileStorage& fs) const
    {
     fs << "{" << "A" << A << "X" << X << "id" << id << "}";
    }

    // 此类的输入
    void read(const FileNode& node)
    {
        A = (int)node["A"];
        X = (double)node["X"];
        id = (string)node["id"];
    }

// 数据成员
public:
    int A;
    double X;
    string id;
};

// 实际输出与输入函数
static void write(FileStorage& fs, const std::string&, const MyData& x)
{
    x.write(fs);
}

static void read(const FileNode& node, MyData& x,
    const MyData& default_value = MyData()){

    if (node.empty())
        x = default_value;
    else
        x.read(node);
}

// FileStorage 输入功能
```

```cpp
static ostream& operator<<(ostream& out, const MyData& m)
{
    out << "{ id = " << m.id << ", ";
    out << "X = " << m.X << ", ";
    out << "A = " << m.A << "}";
    return out;
}

int main(int ac, char** av)
{
    string filename = "C:\\images\\process\\outputfile.yml";

    // 写入
    {
        cout << endl << "写入开始: " << endl;
        Mat R = Mat_<uchar>::eye(3, 3),
            T = Mat_<double>::zeros(3, 1);
        MyData m(1);

        FileStorage fs(filename, FileStorage::WRITE);

        fs << "iterationNr" << 100;

        // 字符串开始
        fs << "strings" << "[";
        fs << "C:\\images\\lena.jpg" << "Awesomeness"
           << "C:\\images\\baboon.jpg";
        // 字符串结束
        fs << "]";

        // 字符串匹配 (mapping)
        fs << "Mapping";
        fs << "{" << "One" << 1;
        fs << "Two" << 2 << "}";

        // cv::Mat
        fs << "R" << R;
        fs << "T" << T;

        // 数据结构
        fs << "MyData" << m;

        // 释放 fs 内存
        fs.release();
        cout << "写入完成." << endl;
    }

    // 读取
    {
        cout << endl << "读取开始: " << endl;
        FileStorage fs;
        fs.open(filename, FileStorage::READ);

        int itNr;
        //fs["iterationNr"] >> itNr;
```

```cpp
        itNr = (int)fs["iterationNr"];
        cout << itNr;
        if (!fs.isOpened())
        {
            cerr << "Failed to open " << filename << endl;
            return 1;
        }

        // 读取字符串并取得节点
        FileNode n = fs["strings"];
        if (n.type() != FileNode::SEQ)
        {
            cerr << "错误，字符串不是循序的！" << endl;
            return 1;
        }

        // Go through the node
        FileNodeIterator it = n.begin(), it_end = n.end();
        for (; it != it_end; ++it)
            cout << (string)*it << endl;

        // Read mappings from a sequence
        n = fs["Mapping"];
        cout << "Two  " << (int)(n["Two"]) << "; ";
        cout << "One  " << (int)(n["One"]) << endl << endl;

        MyData m;
        Mat R, T;

        // Read cv::Mat
        fs["R"] >> R;
        fs["T"] >> T;
        // 读取自己的数据结构
        fs["MyData"] >> m;

        cout << endl
            << "R = " << R << endl;
        cout << "T = " << T << endl << endl;
        cout << "MyData = " << endl << m << endl << endl;

        // 对不存在的节点显示基本特性
        cout << "尝试读取不存在的";
        fs["NonExisting"] >> m;
        cout << endl << "NonExisting = " << endl << m << endl;
    }

    cout << endl
        << "提醒：请使用文字编辑器打开存储的文件 " << filename
        << endl;

    getchar();

    return 0;
}
```

程序说明

1. FileStorage (const string& source, int flags, const string& encoding=string()) FileStorage：类别构造函数

（1）source：是要打开的文件名或要读入的字符串。

如果为文件名，则文件后缀决定其格式，文件后缀再加上.gz 就成为压缩文件。

后 缀 名	格 式
.xml	XML 文件
.yml 或.yaml	YAML 文件

（2）flags：指定操作标志。
- FileStorage::READ：读取文件。
- FileStorage::WRITE：要写入的文件。
- FileStorage::APPEND：要附加的文件。
- FileStorage::MEMORY：由 source 读入数据或将数据写入缓存区。

如果指定 FileStorage::WRITE + FileStorage::MEMORY，source 将决定输出文件的格式，例如 mydata.xml 或 mydata.yml 等。

（3）encoding：文件编码规则，目前不支持 UTF-16 XML 编码规则，只能使用 8 位编码。

2. bool FileStorage::open(const string& filename, int flags, const string& encoding=string())：打开文件

（1）filename：文件名或读入数据的文字字符串；如果为文件名，则文件后缀必须是.xml 或.yml（或是.yaml）。

（2）flags：操作模式。

（3）encoding：文件编码规则。

3. template<typename _Tp> FileStorage& operator<<(FileStorage& fs, const _Tp& value)：将数据写入文件存储（file storage）

或是：

template<typename _Tp> FileStorage& operator<<(FileStorage& fs, const vector<_Tp>& vec)

（1）fs：已打开的文件存储用来写入数据。

（2）value：写入文件存储的数据。

（3）vec：写入文件存储的向量值。

4. template<typename _Tp> void operator>>(const FileNode& n, _Tp& value) 由文件存储读取数据

或是：

template<typename _Tp> void operator>>(const FileNode& n, vector<_Tp>& vec)

　　或是：

template<typename _Tp> FileNodeIterator& operator>>(FileNodeIterator& it, _Tp& value)

　　或是：

template<typename _Tp> FileNodeIterator& operator>>(FileNodeIterator& it, vector<_Tp>& vec)

(1) n：要读取数据的节点（node）。
(2) value：从文件存储读入的数据。
(3) vec：从文件存储读入的向量值。
(4) it：读入数据的迭代器（iterator）。

本程序是展示 OpenCV 处理文字文件的功能，尝试将 filename 内容改成下列不同名称的结果。执行结果如图 2-28 所示。

图 2-28

产生文件的内容如下：

```
outputfile.yml.gz
outputfile.yml

%YAML:1.0
iterationNr: 100
strings:
   - "C:\\images\\lena.jpg"
   - Awesomeness
   - "C:\\images\\baboon.jpg"
Mapping:
   One: 1
   Two: 2
```

```
R: !!opencv-matrix
   rows: 3
   cols: 3
   dt: u
   data: [ 1, 0, 0, 0, 1, 0, 0, 0, 1 ]
T: !!opencv-matrix
   rows: 3
   cols: 1
   dt: d
   data: [ 0., 0., 0. ]
MyData:
   A: 97
   X: 3.1415926535897931e+000
   id: mydata1234
```

产生文件的内容如下:

```
outputfile.xml.gz
outputfile.xml

<?xml version="1.0"?>
<opencv_storage>
<iterationNr>100</iterationNr>
<strings>
  C:\images\lena.jpg Awesomeness C:\images\baboon.jpg</strings>
<Mapping>
  <One>1</One>
  <Two>2</Two></Mapping>
<R type_id="opencv-matrix">
  <rows>3</rows>
  <cols>3</cols>
  <dt>u</dt>
  <data>
    1 0 0 0 1 0 0 0 1</data></R>
<T type_id="opencv-matrix">
  <rows>3</rows>
  <cols>1</cols>
  <dt>d</dt>
  <data>
    0. 0. 0.</data></T>
<MyData>
  <A>97</A>
  <X>3.1415926535897931e+000</X>
  <id>mydata1234</id></MyData>
</opencv_storage>
```

写入

写入的代码如下:

```cpp
#include "opencv2/opencv.hpp"
#include <time.h>

using namespace cv;

int main(int, char** argv)
{
```

```
        FileStorage fs("c:/images/process/test.yml",
                FileStorage::WRITE);

        fs << "frameCount" << 5;
        time_t rawtime; time(&rawtime);
        fs << "calibrationDate" << asctime(localtime(&rawtime));
        Mat cameraMatrix = (Mat_<double>(3, 3)
            << 1000, 0, 320, 0, 1000, 240, 0, 0, 1);
        Mat distCoeffs = (Mat_<double>(5, 1)
            << 0.1, 0.01, -0.001, 0, 0);
        fs << "cameraMatrix" << cameraMatrix
            << "distCoeffs" << distCoeffs;
        fs << "features" << "[";
        for (int i = 0; i < 3; i++)
        {
            int x = rand() % 640;
            int y = rand() % 480;
            uchar lbp = rand() % 256;

            fs << "{:" << "x" << x << "y" << y << "lbp" << "[:";
            for (int j = 0; j < 8; j++)
                fs << ((lbp >> j) & 1);
            fs << "]" << "}";
        }
        fs << "]";
        fs.release();
        return 0;
}
```

代码如图 2-29 所示。

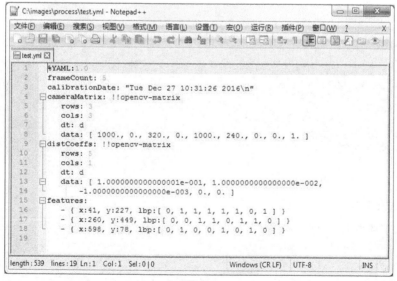

图 2-29

读取

读取的代码如下:

```cpp
#include "opencv2/opencv.hpp"
#include <iostream>

using namespace cv;
using namespace std;

int main(int, char** argv)
{
    FileStorage fs2("c:/images/process/test.yml",
        FileStorage::READ);

    // first method: use (type) operator on FileNode.
    int frameCount = (int)fs2["frameCount"];

    std::string date;
    // second method: use FileNode::operator >>
    fs2["calibrationDate"] >> date;

    Mat cameraMatrix2, distCoeffs2;
    fs2["cameraMatrix"] >> cameraMatrix2;
    fs2["distCoeffs"] >> distCoeffs2;

    cout << "frameCount: " << frameCount << endl
        << "calibration date: " << date << endl
        << "camera matrix: " << cameraMatrix2 << endl
        << "distortion coeffs: " << distCoeffs2 << endl;

    FileNode features = fs2["features"];
    FileNodeIterator it = features.begin(),
        it_end = features.end();

    int idx = 0;
    vector<uchar> lbpval;

    for (; it != it_end; ++it, idx++)
    {
        cout << "feature #" << idx << ": ";
        cout << "x=" << (int)(*it)["x"]
            << ", y=" << (int)(*it)["y"]
            << ", lbp: (";

        (*it)["lbp"] >> lbpval;
        for (int i = 0; i < (int)lbpval.size(); i++)
            cout << " " << (int)lbpval[i];

        cout << ")" << endl;
    }
    fs2.release();
    getchar();
}
```

程序说明

1. FileNodeIterator FileNode::begin() const：返回指向第一个节点元素的迭代器。
2. FileNodeIterator FileNode::end() const：返回指向最后节点元素的迭代器。

执行结果如图 2-30 所示。

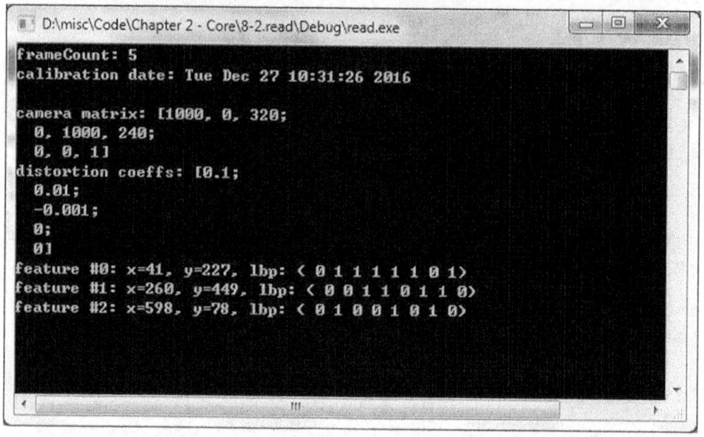

图 2-30

2.9 与 OpenCV 1 互通

本节范例是说明 OpenCV 1 与 OpenCV 2 指令的不同以及如何共用图文件数据。因为许多较早的图书使用 OpenCV 1，所以读者看完本程序了解 OpenCV 1 程序的用法后，也知道如何将 OpenCV 1 程序转换成 OpenCV 2。

```
#include <stdio.h>
#include <iostream>

#include <opencv2/core/core.hpp>
#include <opencv2/imgproc/imgproc.hpp>
#include <opencv2/highgui/highgui.hpp>

using namespace cv;
using namespace std;

// 使用 OpenCV 2 则将下行注释掉
#define DEMO_MIXED_API_USE

int main( int argc, char** argv )
{
    const char* imagename = argc > 1 ? argv[1] : "C:\\images\\lena.jpg";
```

```cpp
// OpenCV 1
#ifdef DEMO_MIXED_API_USE

    Ptr<IplImage> IplI = cvLoadImage(imagename);
    if(IplI.empty())
    {
        cerr << "无法载入图文件 " <<  imagename << endl;
        return -1;
    }

    // 数据转成 2.0 格式
    Mat I(IplI);
#else
    // cvLoadImage 的新用法
    Mat I = imread(imagename);

    // 也可以使用 if( !I.data )
    if(I.empty())
    {
        cerr << "无法载入图文件 " <<  imagename << endl;
        return -1;
    }
#endif

    // 将图转成 YUV 颜色空间
    Mat I_YUV;
    cvtColor(I, I_YUV, COLOR_BGR2YCrCb);

    // 使用 STL 的向量结构存储多个 Mat 对象
    vector<Mat> planes;

    // 将图分成各自的色面 (Y U V)
    split(I_YUV, planes);

#if 1
    // 扫描 Mat
    // 方法 1: 使用迭代器处理 Y 平面
    MatIterator_<uchar> it = planes[0].begin<uchar>(),
            it_end = planes[0].end<uchar>();

    for(; it != it_end; ++it)
    {
        double v = *it * 1.7 + rand()%21 - 10;
        *it = saturate_cast<uchar>(v*v/255);
    }

    for( int y = 0; y < I_YUV.rows; y++ )
    {
        // 方法 2: 使用原始点 ( row pointer ) 处理第一个色度面 ( chroma plane )
        uchar* Uptr = planes[1].ptr<uchar>(y);
        for( int x = 0; x < I_YUV.cols; x++ )
        {
            Uptr[x] = saturate_cast<uchar>((Uptr[x]-128)/2 + 128);

            // 方法 3: 使用各自原素的访问处理第二个色度面
```

```cpp
                uchar& Vxy = planes[2].at<uchar>(y, x);
                Vxy = saturate_cast<uchar>((Vxy-128)/2 + 128);
            }
        }

#else
    // 使用指定的大小与格式建立矩阵
        Mat noisyI(I.size(), CV_8U);

        // 使用正常分布的随机值填满矩阵
        // 也可以使用一致分布的随机值 randu()
        randn(noisyI, Scalar::all(128), Scalar::all(20));

        // 将 noisyI 弄模糊,核心大小是 3x3,两个 sigma 值设为 0.5
        GaussianBlur(noisyI, noisyI, Size(3, 3), 0.5, 0.5);

        const double brightness_gain = 0;
        const double contrast_gain = 1.7;

#ifdef DEMO_MIXED_API_USE

        IplImage cv_planes_0 = planes[0], cv_noise = noisyI;
        cvAddWeighted(&cv_planes_0, contrast_gain, &cv_noise,
                1, -128 + brightness_gain, &cv_planes_0);
#else
        addWeighted(planes[0], contrast_gain, noisyI, 1,
                -128 + brightness_gain, planes[0]);
#endif

        const double color_scale = 0.5;

        // Mat::convertTo() 取代 cvConvertScale.
        // 必须明确指定输出矩阵的格式(保留原先的格式 - planes[1].type())
        planes[1].convertTo(planes[1], planes[1].type(),
                color_scale, 128*(1-color_scale));

        // cv::convertScale 的另一种形式
        planes[2] = Mat_<uchar>(planes[2]*color_scale + 128*(1-color_scale));

        // Mat::mul 取代 cvMul().
        planes[0] = planes[0].mul(planes[0], 1./255);
#endif

        // 合并结果
        merge(planes, I_YUV);

        // 产生 RGB 的图
        cvtColor(I_YUV, I, CV_YCrCb2BGR);

        namedWindow("有粒纹的图", WINDOW_AUTOSIZE);
#ifdef DEMO_MIXED_API_USE
        // 这是示范 I 与 IplI 真的共享数据
        // 所以存在 I,也会存在 IplI
        cvShowImage("有粒纹的图", IplI);
```

```
#else
    // 用新的 MATLAB 形态显示
    imshow("有粒纹的图", I);
#endif
    waitKey();

    // 所有内存会被 Vector<>、 Mat 与 Ptr<> 的解构函数释放
    return 0;
}
```

执行结果如图 2-31 所示。

（a）#if 0（模糊与噪声的）　　　　　　　（b）#if 1（无模糊与噪声的）

图 2-31

第 3 章
HighGUI 模块

HighGUI 的全名是"高端用户图形界面"（High-level Graphical User Interface）。这个模块主要是与硬件输入/输出设备（例如相机、键盘与鼠标），以及系统（包括操作系统、文件系统）通信。所以，打开窗口，读取图像（静态图或动态视频）与显示图像，是这个模块的功能，也可以说 OpenCV 的基石就是 Core 与 HighGUI 两个模块。

虽然这个模块取名 GUI，却只提供将图形文件或视频显示在屏幕上的简单功能，最多只是在显示的窗口上增加窗口显示的滑块（Track-Bar），硬件界面也只是鼠标与键盘。这是当初在设计 OpenCV 的时候，希望 OpenCV 可以适用于各种系统环境。因为各系统处理 GUI 的方式不同、硬件界面不同，所以只提供简单必要的图像与视频的显示。其他针对所使用系统的 GUI，设计者自行在该系统中去设计。

关于 GUI 部分，如果在 Windows 的环境使用 C++开发，可以使用 MFC 或 C++/CLI。C++/CLI 能够让 C++语言使用.NET 功能，关于使用 MFC 的范例可参考相关技术文章：http://www.codeproject.com/Articles/741055/Video-Capture-using-OpenCV-with-VCplusplus。如果使用 C#开发，那就有 Emgu CV 或 AForge.NET 可供使用，下载的网址分别是 http://sourceforge.net/projects/emgucv/和 http://www.aforgenet.com/。

读者有兴趣可以在下列网站搜索到所有与 OpenCV 相关的软件：http://sourceforge.net/directory/os/windows/freshness:recently-updated/?q=opencv。

3.1 滑块功能

实现滑块功能的代码如下：

```
#include "opencv2/highgui/highgui.hpp"
#include <stdio.h>
using namespace cv;
/** 声明全局变量 */
const int alpha_slider_max = 100;
int alpha_slider;
double alpha;
double beta;
/** 声明图像数组 */
Mat src1;
Mat src2;
Mat dst;
static void on_trackbar(int, void*)
{
    alpha = (double)alpha_slider / alpha_slider_max;
    beta = (1.0 - alpha);
    addWeighted(src1, alpha, src2, beta, 0.0, dst);
    imshow("Linear Blend", dst);
}
int main(void)
{
    /// 读取同样大小与类型的两个图形文件
```

```
src1 = imread("C:\\images\\LinuxLogo.jpg");
src2 = imread("C:\\images\\WindowsLogo.jpg");
if (!src1.data) { printf("Error loading src1 \n"); return -1; }
if (!src2.data) { printf("Error loading src2 \n"); return -1; }
/// 起始值
alpha_slider = 0;
/// 建立窗口
namedWindow("Linear Blend", 1);
char TrackbarName[50];
sprintf(TrackbarName, "Alpha %d", alpha_slider_max);
/// 产生滑块回复函数
createTrackbar(TrackbarName, "Linear Blend", &alpha_slider,
    alpha_slider_max, on_trackbar);
/// 等待按键
waitKey(0);
return 0;
}
```

执行结果如图 3-1 所示。

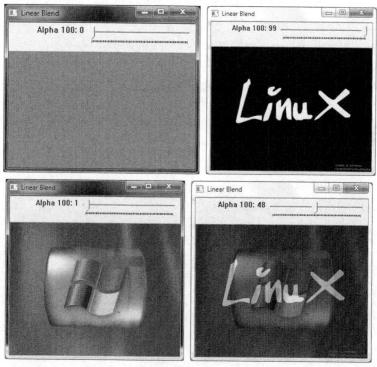

图 3-1

程序说明

CreateTrackbar (const string& trackbarname, const string& winname, int* value, int count, TrackbarCallback onChange=0, void* userdata=0)：建立窗口滑块回调（callback）功能

（1）trackbarname：滑块名称。

（2）winname：滑块要呈现的窗口名称。
（3）value：滑块所在位置的值。
（4）count：滑块允许的最大值，最小值为 0。
（5）onChange：滑块值变动实际处理的函数，此函数第一个参数是滑块位置值，第二个参数是 userdata。
（6）userdata-要传入处理滑块函数的值。

滑块的作用是许多设置通过此功能调整，当我们认为可以时，滑杆的值就是将来放到程序中的值。这种在某种环境下调整找到最佳值的过程，在视频处理中称为"学习"。

3.2 读取视频文件进行相似性比较

本范例介绍的是视频文件的处理，而关于即时视频的拍摄将在第 7 章讨论。本节与第 7 章内容表明了：在 OpenCV 中，这两种处理方式是完全一样的，所以读者也可以先看完第 7 章再回来阅读本小节。

视频（video）就是由连续的图像（image）组成，组成视频的图像称为帧（frame）。视频播放时有帧速率（frame rate），也就是播放时两个帧之间的时间，而相机每秒能拍摄的帧数是有限制的。

本小节程序是介绍用峰值信噪比（PSNR）与结构相似性图像质量指标（SSIM）来检查两个视频文件的相似度，同时看看用新压缩法的视频文件与原视频的差异。其代码如下：

```
#include <iostream>
#include <string>
#include <iomanip>    // 用于控制小数点显示的精确度
#include <sstream>    // 字符串转换成数字
#include <opencv2/core/core.hpp>
#include <opencv2/imgproc/imgproc.hpp>
#include <opencv2/highgui/highgui.hpp>
using namespace std;
using namespace cv;
// 函数声明
double getPSNR(const Mat& I1, const Mat& I2);
Scalar getMSSIM(const Mat& I1, const Mat& I2);
int main(int argc, char *argv[])
{
    // 指定视频
    argv[1] = "C:\\images\\Megamind.avi";
    // 比较的视频
    argv[2] = "C:\\images\\Megamind_bugy.avi";
    // PSNR 触发值
    argv[3] = "350";
    // 帧之间的等待时间
    // 用来进行两视频间差异的人为比较
    argv[4] = "10";
```

```cpp
    stringstream conv;
    const string sourceReference = argv[1],
          sourceCompareWith = argv[2];
    int psnrTriggerValue, delay;
    conv << argv[3] << endl << argv[4];
    // psnrTriggerValue = argv[3]
    // delay = argv[4]
    conv >> psnrTriggerValue >> delay;
    char c;
    // 视频起点
    int frameNum = -1;
    VideoCapture captRefrnc(sourceReference),
                 captUndTst(sourceCompareWith);
    if (!captRefrnc.isOpened())
    {
       cout << "无法打开指定视频 " << sourceReference << endl;
       return -1;
    }
    if (!captUndTst.isOpened())
    {
       cout << "无法打开比较的视频 " << sourceCompareWith << endl;
       return -1;
    }
    // 比较两个图像文件大小
    Size refS = Size((int)captRefrnc.get(CV_CAP_PROP_FRAME_WIDTH),
                    (int)captRefrnc.get(CV_CAP_PROP_FRAME_HEIGHT));
    Size uTSi = Size((int)captUndTst.get(CV_CAP_PROP_FRAME_WIDTH),
                    (int)captUndTst.get(CV_CAP_PROP_FRAME_HEIGHT));
    if (refS != uTSi)
    {
       cout << "两个视频不同大小!!! 结束" << endl;
       return -1;
    }
    const char* WIN_UT = "比较的视频";
    const char* WIN_RF = "指定视频";
    // 建立窗口
    namedWindow(WIN_RF, CV_WINDOW_AUTOSIZE);
    namedWindow(WIN_UT, CV_WINDOW_AUTOSIZE);
    // 将窗口放置特定位置
    moveWindow(WIN_RF, 0, 100);
    moveWindow(WIN_UT, -10 + refS.width + 10, 100);
    cout << "参考视频分辨率: 宽 ="
         << refS.width << " 高 =" << refS.height
         << " 第 " << captRefrnc.get(CV_CAP_PROP_FRAME_COUNT)
         << " 个帧" << endl;
    cout << "PSNR 触发值 " << setiosflags(ios::fixed)
         << setprecision(3) << psnrTriggerValue << endl;
    Mat frameReference, frameUnderTest;
    double psnrV;
    Scalar mssimV;
    // 显示结果
    for (;;)
    {
       captRefrnc >> frameReference;
       captUndTst >> frameUnderTest;
```

```cpp
        if (frameReference.empty() || frameUnderTest.empty())
        {
            cout << " < < <   结束!   > > > ";
            break;
        }
        ++frameNum;
        cout << "Frame: " << frameNum << "# ";
        ///////////////////// PSNR /////////////////////////////
        psnrV = getPSNR(frameReference, frameUnderTest);
        cout << setiosflags(ios::fixed)
             << setprecision(3) << psnrV << "dB";
        ///////////////////// MSSIM ////////////////////////////
        if (psnrV < psnrTriggerValue && psnrV)
        {
            mssimV = getMSSIM(frameReference, frameUnderTest);
            cout << " MSSIM: "
                 << " R " << setiosflags(ios::fixed) << setprecision(2)
                 << mssimV.val[2] * 100 << "%"
                 << " G " << setiosflags(ios::fixed) << setprecision(2)
                 << mssimV.val[1] * 100 << "%"
                 << " B " << setiosflags(ios::fixed) << setprecision(2)
                 << mssimV.val[0] * 100 << "%";
        }
        cout << endl;
        ///////////////////// Show Image ////////////////////
        imshow(WIN_RF, frameReference);
        imshow(WIN_UT, frameUnderTest);
        c = (char)cvWaitKey(delay);
        if (c == 27) break;
    }
    return 0;
}
double getPSNR(const Mat& I1, const Mat& I2)
{
    Mat s1;
    // |I1 - I2|
    absdiff(I1, I2, s1);
    // 将CV_32F转换成Mat,因为8 bits不能平方
    s1.convertTo(s1, CV_32F);
    // |I1 - I2|^2
    s1 = s1.mul(s1);
    // sum elements per channel
    Scalar s = sum(s1);
    // 通道加和
    double sse = s.val[0] + s.val[1] + s.val[2];
    // 若值太小就传回0
    if (sse <= 1e-10)
        return 0;
    else
    {
        double mse = sse / (double)(I1.channels() * I1.total());
        double psnr = 10.0 * log10((255 * 255) / mse);
        return psnr;
    }
}
```

```
Scalar getMSSIM(const Mat& i1, const Mat& i2)
{
    const double C1 = 6.5025, C2 = 58.5225;
    /*************** INITS ********************/
    int d = CV_32F;
    Mat I1, I2;
    // 将 CV_32F 转换成 Mat，因为 8 bits 不能平方
    i1.convertTo(I1, d);
    i2.convertTo(I2, d);
    // I2^2
    Mat I2_2 = I2.mul(I2);

    // I1^2
    Mat I1_2 = I1.mul(I1);

    // I1 * I2
    Mat I1_I2 = I1.mul(I2);
    /*********** END INITS ******************/
    Mat mu1, mu2;
    // 开始计算
    GaussianBlur(I1, mu1, Size(11, 11), 1.5);
    GaussianBlur(I2, mu2, Size(11, 11), 1.5);
    Mat mu1_2 = mu1.mul(mu1);
    Mat mu2_2 = mu2.mul(mu2);
    Mat mu1_mu2 = mu1.mul(mu2);
    Mat sigma1_2, sigma2_2, sigma12;
    GaussianBlur(I1_2, sigma1_2, Size(11, 11), 1.5);
    sigma1_2 -= mu1_2;
    GaussianBlur(I2_2, sigma2_2, Size(11, 11), 1.5);
    sigma2_2 -= mu2_2;
    GaussianBlur(I1_I2, sigma12, Size(11, 11), 1.5);
    sigma12 -= mu1_mu2;
    //////////////// FORMULA ////////////////
    Mat t1, t2, t3;
    t1 = 2 * mu1_mu2 + C1;
    t2 = 2 * sigma12 + C2;
    // t3 = ((2*mu1_mu2 + C1).*(2*sigma12 + C2))
    t3 = t1.mul(t2);
    t1 = mu1_2 + mu2_2 + C1;
    t2 = sigma1_2 + sigma2_2 + C2;
    // t1 =((mu1_2 + mu2_2 + C1).*(sigma1_2 + sigma2_2 + C2))
    t1 = t1.mul(t2);
    Mat ssim_map;
    // ssim_map =  t3./t1;
    divide(t3, t1, ssim_map);
    // mssim = ssim_map 的平均值
    Scalar mssim = mean(ssim_map);
    return mssim;
}
```

程序说明

1. absdiff (InputArray src1, InputArray src2, OutputArray dst)：**计算两图差异的绝对值**

    （1）src1：第一个输入图像或 Scalar 颜色值。

（2）src2：第二个输入图像或 Scalar 颜色值。

（3）dst：差异结果图像，大小和类型与输入图像相同。

2. GaussianBlur (InputArray src, OutputArray dst, Size ksize, double sigmaX, double sigmaY=0, int borderType=BORDER_DEFAULT)：用高斯滤波器（filter）平滑图像

（1）src：输入图像。

（2）dst：输出图像。

（3）ksize：高斯核心值的大小（kernel size）。

（4）sigmaX：高斯 X 轴方向核心值标准差。

（5）sigmaY：高斯 Y 轴方向核心值标准差，如果值为 0 就与 sigmaX 相同。

（6）borderType：像素外推法（extrapolation method）边界的类型。

执行结果如图 3-2 所示。

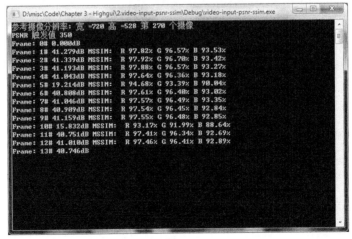

图 3-2

从上面的数据来看第 5 帧是有问题的,PSNR 值只有 19.214dB,而 RGB 值在 MSSIM 也偏低,也就是视频中显示的白线瑕疵。

3.3 产生视频文件

视频文件一般除了视频压缩与压缩的编码规格之外,还有音轨与字幕资料。而 OpenCV 只是视频处理的程序库,当初在开发的时候为求简化,并没有纳入音轨与字幕资料部分。但是 OpenCV 还是可以处理含有这些资料的视频文件。

现在所有视频编码规格都有独特的短名,而最长 4 个字符。例如 XVID、CIVX 和 H264,这就是所谓的四字字符编码(four character code)。OpenCV 使用 get 函数取得编码规格,get 返回 double 类型的值,这个值是 64 位的,但是 4 个字符只有 32 位,所以编码规格式存在 64 位中的左边的 32 位,也就是 64 位中较低的位。要抛弃较高的位,最简单方式就是在 C 语言中强制转型为 int,所以程序都会做此处理:static_cast<int>VideoCapture.get()。

```
#include <iostream>
#include <string>
#include <opencv2/core/core.hpp>
#include <opencv2/highgui/highgui.hpp>
using namespace std;
using namespace cv;
int main(int argc, char *argv[])
{
   argv[1] = "C:\\images\\Megamind.avi";
   argv[2] = "R";
   argv[3] = "Y";
   // 要读取的文件名
   const string source = argv[1];
   // 使用输入文件名的编码格式 (codec type)
   const bool askOutputType = argv[3][0] == 'Y';
   // 打开文件
   VideoCapture inputVideo(source);
   if (!inputVideo.isOpened())
   {
      cout << "无法打开文件: " << source << endl;
      return -1;
   }
   // 寻找文件后缀名的点'.'
   string::size_type pAt = source.find_last_of('.');
   // 原程序生成文件名在 argv[1]视频文件存放的位置,并将文件名改为
   // C:\\images\\MegamindR.avi
   // const string NAME = source.substr(0, pAt) + argv[2][0] + ".avi";
   // 更改程序将文件放到指定位置
   const string NAME = "C:\\images\\process\\Megamind.avi";
   // 取得输入文件编码格式
   int ex = static_cast<int>(inputVideo.get(CV_CAP_PROP_FOURCC));
   // 通过位的运算,将 int 转换为 char
```

```cpp
char EXT[] = { (char)(ex & 0XFF), (char)((ex & 0XFF00) >> 8),
    (char)((ex & 0XFF0000) >> 16), (char)((ex & 0XFF000000) >> 24), 0 };
// 取得输入视频文件大小
Size S = Size((int)inputVideo.get(CV_CAP_PROP_FRAME_WIDTH),
    (int)inputVideo.get(CV_CAP_PROP_FRAME_HEIGHT));
// 打开输出文件
VideoWriter outputVideo;
if (askOutputType)
    outputVideo.open(NAME, ex = -1,
        inputVideo.get(CV_CAP_PROP_FPS), S, true);
else
    outputVideo.open(NAME, ex,
        inputVideo.get(CV_CAP_PROP_FPS), S, true);
if (!outputVideo.isOpened())
{
    cout << "无法打开输出文件供写入: " << source << endl;
    getchar();
    return -1;
}
cout << "输入视频分辨率的: 宽=" << S.width
    << "  高=" << S.height << " of nr#: "
    << inputVideo.get(CV_CAP_PROP_FRAME_COUNT)
    << endl;
cout << "输入的编码格式(codec type): " << EXT << endl;
// 选择存储的通道
int channel = 2;
switch (argv[2][0])
{
    case 'R':
        channel = 2;
        break;
    case 'G':
        channel = 1;
        break;
    case 'B':
        channel = 0;
        break;
}
Mat src, res;
vector<Mat> spl;
for (;;)
{
    // 读取文件
    inputVideo >> src;
    // 检查是否结束
    if (src.empty())
        break;
    // 只取得我们要的通道
    split(src, spl);
    for (int i = 0; i < 3; ++i)
        if (i != channel)
            spl[i] = Mat::zeros(S, spl[0].type());
    merge(spl, res);
    // 输出文件
    outputVideo << res;
}
cout << "结束输出" << endl;
```

```
    getchar();
    return 0;
}
```

执行结果

执行此程序之后,会弹出选择安装在计算机内所有的编码规格,如图 3-3 所示。请选择如下设置,如图 3-4 所示。

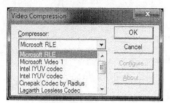

图 3-3

图 3-4

更改程序不同的通道值,得到如图 3-5 所示的结果。

(a) 彩色原图

(b) 红色图

(c) 绿色图

(d) 蓝色图

图 3-5

第 4 章 ImgProc 模块

这个模块主要是进行图像处理,所谓图像处理就是使用图像结构内定义的高级运算来完成工作。

4.1 图像的平滑化

平滑化(smoothing)也称为模糊化,经常用来进行简单的图像处理。其作用有许多,但本例只介绍线性滤波器(linear filter)以减少噪声(noise),平滑化也是降低图像分辨率很重要的方法。

```
#include <iostream>
#include <vector>
#include "opencv2/imgproc/imgproc.hpp"
#include "opencv2/highgui/highgui.hpp"
#include "opencv2/features2d/features2d.hpp"
using namespace std;
using namespace cv;
/// 声明全局变量
int DELAY_CAPTION = 1500;
int DELAY_BLUR = 100;
int MAX_KERNEL_LENGTH = 31;
Mat src; Mat dst;
char window_name[] = "Smoothing Demo";
/// 声明函数
int display_caption(const char* caption);
int display_dst(int delay);
int main(void)
{
    namedWindow(window_name, WINDOW_AUTOSIZE);
    /// 加载图形文件
    src = imread("C:\\images\\lena.jpg", 1);
    if (display_caption("Original Image") != 0)
        return 0;
    dst = src.clone();
    if (display_dst(DELAY_CAPTION) != 0)
        return 0;
    /// 使用 Homogeneous 模糊法
    if (display_caption("Homogeneous Blur") != 0)
        return 0;
    for (int i = 1; i < MAX_KERNEL_LENGTH; i = i + 2)
    {
        blur(src, dst, Size(i, i), Point(-1, -1));
        if (display_dst(DELAY_BLUR) != 0)
            return 0;
    }
    /// 使用 Gaussian 模糊法
    if (display_caption("Gaussian Blur") != 0)
```

```cpp
        return 0;
    for (int i = 1; i < MAX_KERNEL_LENGTH; i = i + 2)
    {
        GaussianBlur(src, dst, Size(i, i), 0, 0);
        if (display_dst(DELAY_BLUR) != 0)
            return 0;
    }
    /// 使用 Median 模糊法
    if (display_caption("Median Blur") != 0)
        return 0;
    for (int i = 1; i < MAX_KERNEL_LENGTH; i = i + 2)
    {
        medianBlur(src, dst, i);
        if (display_dst(DELAY_BLUR) != 0)
        { return 0; }
    }
    /// 使用 Bilateral 模糊法
    if (display_caption("Bilateral Blur") != 0)
        return 0;
    for (int i = 1; i < MAX_KERNEL_LENGTH; i = i + 2)
    {
        bilateralFilter(src, dst, i, i * 2, i / 2);
        if (display_dst(DELAY_BLUR) != 0)
            return 0;
    }
    /// 显示结束并等待按键
    display_caption("End: Press a key!");
    waitKey(0);
    return 0;
}
/// 显示文字
int display_caption(const char* caption)
{
    dst = Mat::zeros(src.size(), src.type());
    putText(dst, caption,
        Point(src.cols / 4, src.rows / 2),
        FONT_HERSHEY_COMPLEX, 1, Scalar(255, 255, 255));
    imshow(window_name, dst);
    int c = waitKey(DELAY_CAPTION);
    if (c >= 0) { return -1; }
    return 0;
}
/// 显示图像
int display_dst(int delay)
{
    imshow(window_name, dst);
    int c = waitKey(delay);
    if (c >= 0) { return -1; }
    return 0;
}
```

程序说明

1. blur (InputArray src, OutputArray dst, Size ksize, Point anchor=Point(-1, -1), int borderType=BORDER_DEFAULT)：归一化方格滤波器来平滑图像

   （1）src：输入图像。
   （2）dst：输出图像，大小和类型与输入图像相同。
   （3）ksize：模糊化的核大小。
   （4）anchor：锚点，默认值是 Point(-1, -1)，锚点就是核中点。
   （5）borderType：像素外推法边缘的类型。

2. medianBlur (InputArray src, OutputArray dst, int ksize)：使用平均滤波器模糊图像

   （1）src：输入图像。
   （2）dst：输出图像；大小和类型与输入图像相同。
   （3）ksize：光圈线性大小（aperture linear size），必须是大于 1 的奇数。

3. bilateralFilter (InputArray src, OutputArray dst, int d, double sigmaColor, double sigmaSpace, int borderType=BORDER_DEFAULT)：左右对称滤波图像

   （1）src：输入图像。
   （2）dst：输出图像；大小和类型与输入图像相同。
   （3）d：每个像素临近地区的直径。
   （4）sigmaColor：颜色空间（color space）内的滤波器标准差（filter sigma）。
   （5）sigmaSpace：坐标空间（coordinate space）内的滤波器标准差。
   （6）borderType：像素外推法边缘的类型。

   各方法的执行结果如图 4-1 所示。

图 4-1

图 4-1（续）

图 4-1（续）

4.2 腐蚀与膨胀

一般图像处理是针对图像做形状的改变，而形态处理（morphological）则是对图像进行结构性的改变，而最基本的形态处理就是腐蚀（Erosion）与膨胀（Dilation）。

形态处理的应用有以下几种。
- 删除噪声。
- 图中个别原素的隔离与不同原素的连接。
- 寻找图中浓密点（intensity bump）与空洞（hole）。

如果原图是：

经过膨胀处理后变成下图，也就是背景（白色部分）膨胀了。

经过腐蚀处理后变成下图，也就是背景被腐蚀了。

具体代码如下:

```cpp
#include "opencv2/imgproc/imgproc.hpp"
#include "opencv2/highgui/highgui.hpp"
#include <stdlib.h>
#include <stdio.h>
using namespace cv;
/// 声明全局变量
Mat src, erosion_dst, dilation_dst;
char text[] = "Element: 0:Rect 1:Cross 2:Ellipse";
int erosion_elem = 0;
int erosion_size = 0;
int dilation_elem = 0;
int dilation_size = 0;
int const max_elem = 2;
int const max_kernel_size = 21;
/** 声明函数 */
void Erosion(int, void*);
void Dilation(int, void*);
int main(int, char** argv)
{
    /// 加载图形文件
    src = imread("C:\\images\\opencv-logo.png");
    if (!src.data)
        return -1;
    /// 建立窗口
    namedWindow("腐蚀", WINDOW_AUTOSIZE);
    namedWindow("膨胀", WINDOW_AUTOSIZE);
    moveWindow("膨胀", src.cols, 0);
    /// 建立腐蚀(Erosion)滑杆
    createTrackbar("Element:", "腐蚀",
        &erosion_elem, max_elem, Erosion);
    createTrackbar("Kernel:", "腐蚀",
        &erosion_size, max_kernel_size, Erosion);
    /// 建立膨胀(Dilation)滑杆
    createTrackbar("Element:", "膨胀",
        &dilation_elem, max_elem, Dilation);
    createTrackbar("Kernel:", "膨胀",
        &dilation_size, max_kernel_size, Dilation);
    /// 主运算开始
    Erosion(0, 0);
    Dilation(0, 0);
    waitKey(0);
    return 0;
}
void Erosion(int, void*)
{
    int erosion_type = 0;
    if (erosion_elem == 0) {
        erosion_type = MORPH_RECT;
    } else if (erosion_elem == 1) {
        erosion_type = MORPH_CROSS;
    } else if (erosion_elem == 2) {
        erosion_type = MORPH_ELLIPSE;
    }
    Mat element = getStructuringElement(erosion_type,
```

```
      Size(2 * erosion_size + 1, 2 * erosion_size + 1),
      Point(erosion_size, erosion_size));
   /// 使用腐蚀(erosion)运算
   erode(src, erosion_dst, element);
   // 取得文字中心点
   Point textOrg(10, 25);
   putText(erosion_dst, text, textOrg, FONT_HERSHEY_SIMPLEX, 1, 3);
   imshow("腐蚀", erosion_dst);
}
void Dilation(int, void*)
{
   int dilation_type = 0;
   if (dilation_elem == 0) {
      dilation_type = MORPH_RECT;
   } else if (dilation_elem == 1) {
      dilation_type = MORPH_CROSS;
   } else if (dilation_elem == 2) {
      dilation_type = MORPH_ELLIPSE;
   }
   Mat element = getStructuringElement(dilation_type,
      Size(2 * dilation_size + 1, 2 * dilation_size + 1),
      Point(dilation_size, dilation_size));
   /// 使用膨胀(dilation)运算
   dilate(src, dilation_dst, element);
   // 取得文字中心点
   Point textOrg(10, 25);
   putText(dilation_dst, text, textOrg, FONT_HERSHEY_SIMPLEX, 1, 3);
   imshow("膨胀", dilation_dst);
}
```

程序说明

1. Mat getStructuringElement(int shape, Size ksize, Point anchor=Point(-1, -1))：返回结构性元素(structuring element)

　　（1）shape：元素形状。
- MORPH_RECT：长方型结构性元素。
- MORPH_ELLIPSE：椭圆形结构性元素。
- MORPH_CROSS：十字形结构性元素。
- CV_SHAPE_CUSTOM：自定义结构性元素。

　　（2）ksize：结构性元素的大小。

　　（3）anchor：结构性元素内锚点的位置。

2. erode (InputArray src, OutputArray dst, InputArray kernel, Point anchor, int iterations, int borderType, const Scalar& borderValue)：用结构性元素腐蚀图像

　　（1）src：输入图像。

　　（2）dst：输出图像。

　　（3）kernel：用于腐蚀图像的结构性元素。

（4）anchor：锚点。

（5）iterations：腐蚀次数。

（6）borderType：像素外推法边缘的类型。

（7）borderValue：边缘的值。

3. dilate (InputArray src, OutputArray dst, InputArray kernel, Point anchor=Point(-1,-1), int iterations=1, int borderType=BORDER_CONSTANT, const Scalar& borderValue=morphologyDefaultBorderValue())：用结构性元素膨胀图像

（1）src：输入图像。

（2）dst：输出图像。

（3）kernel：用于腐蚀图像的结构性元素。

（4）anchor：锚点。

（5）iterations：膨胀次数。

（6）borderType：像素外推法边缘的类型。

（7）borderValue：边缘的值。

执行结果如图 4-2 所示。

(a) 原图　　　　　　　　　(b) 腐蚀　　　　　　　　　(c) 膨胀

图 4-2

如果滑杆文字太多，将会省略失去文字说明意义，所以作者将要表达的说明显示于图中，此例说明是 Element 有 3 种选项。

如果滑杆内容字有换行符号，使用 Visual Studio 开发是没有换行效果的，这应该是 OpenCV 的错误，因为在别的环境是有换行作用的，也就不会看不到要表达的内容。

4.3 更多形态处理

上一节介绍了两种基本形态处理，组合处理两种运算将产生更多图的变化。

开放（Opening）

- 这是处理完膨胀后再进行腐蚀处理。

  dst=open(src,element)=dilate(erode(src,element))

- 在删除小对象时很好用，例如黑色背景上亮的对象，下方右边的图就是做开放处理后的结果。

封闭（Closing）

- 这是处理完腐蚀后再进行膨胀处理。

  dst=close(src,element)=erode(dilate(src,element))

- 在删除小洞时很好用。

形态渐层处理(Morphological Gradient)

- 这是腐蚀与膨胀相减的处理。

  dst=$morph_{grad}$(src,element)=dilate(src,element)-erode(src,element)

- 在寻找对象的轮廓（outline）时很好用。

上帽（Top Hat）

- 是原图与图进行开放处理后的差。

黑帽（Black Hat）

- 是图进行封闭处理后与原图的差。

具体代码如下：

```
#include "opencv2/imgproc/imgproc.hpp"
#include "opencv2/highgui/highgui.hpp"
#include <stdlib.h>
#include <stdio.h>
```

```cpp
using namespace cv;
/// 声明全局变量
Mat src, dst;
int morph_elem = 0;
int morph_size = 0;
int morph_operator = 0;
int const max_operator = 4;
int const max_elem = 2;
int const max_kernel_size = 21;
const char* window_name = "Morphology Transformations Demo";
/** 声明函数 */
void Morphology_Operations(int, void*);
int main(int, char** argv)
{
    /// 加载图形文件
    src = imread("C:\\images\\baboon.jpg");
    if (!src.data)
        return -1;
    /// 建立窗口
    namedWindow(window_name, CV_WINDOW_AUTOSIZE);
    /// 建立滑杆以选择 Morphology 运算
    createTrackbar("Operator:\n 0: Opening - 1: Closing  \n 2: Gradient - 3: Top Hat \n 4: Black Hat", window_name, &morph_operator, max_operator, Morphology_Operations);
    /// 建立滑杆以选择 kernel 类型
    createTrackbar("Element:\n 0: Rect - 1: Cross - 2: Ellipse", window_name,
        &morph_elem, max_elem,
        Morphology_Operations);
    /// 建立滑杆以选择 kernel 大小
    createTrackbar("Kernel size:\n 2n +1", window_name,
        &morph_size, max_kernel_size,
        Morphology_Operations);
    /// 主运算开始
    Morphology_Operations(0, 0);
    waitKey(0);
    return 0;
}
void Morphology_Operations(int, void*)
{
    // Since MORPH_X : 2,3,4,5 and 6
    int operation = morph_operator + 2;
    Mat element = getStructuringElement(morph_elem,
        Size(2 * morph_size + 1, 2 * morph_size + 1),
        Point(morph_size, morph_size));
    /// 使用特定的 morphology 运算
    morphologyEx(src, dst, operation, element);
    imshow(window_name, dst);
}
```

程序说明

morphologyEx (InputArray src, OutputArray dst, int op, InputArray kernel, Point anchor=Point(-1, -1), int iterations=1, int borderType=BORDER_CONSTANT, const Scalar& borderValue=morphologyDefaultBorderValue())：高级形态变换

（1）src：输入图像。

（2）dst：输出图像。

（3）op：形态运算类型。

- MORPH_OPEN：开放运算。
- MORPH_CLOSE：关闭运算。
- MORPH_GRADIENT：形态倾斜度。
- MORPH_TOPHAT：上帽（top hat）。
- MORPH_BLACKHAT：黑帽（black hat）。

（4）kernel：用于腐蚀图像的结构性元素。

（5）anchor：锚点。

（6）iterations：膨胀次数。

（7）borderType：像素外推法边缘的类型。

（8）borderValue：边缘的值。

执行结果如图 4-3 和图 4-4 所示。

原图

图 4-3

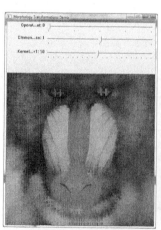

（a）使用滑杆产生开放效果

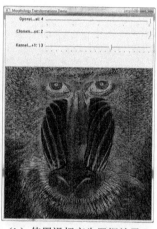

（b）使用滑杆产生黑帽效果

图 4-4

4.4 图像金字塔

在本程序中，我们展示了使用 pryUp 与 pryDown 这两个函数，对图片进行上下取样（Upsample、Downsample）。虽然 OpenCV 的 resize 函数就可以对图做几何变化，但是图像金字塔（image pyramids）较广泛用于各种图像之中。图像金字塔是由许多图像组合而成，由原图相继向下取样，直到上升到希望的点才停止。

图像金字塔有两种，即 Gaussian 金字塔和 Laplacian 金字塔。Gaussian 金字塔，是有很多层而越往上面越小，每上一层就缩小成原来的 1/4，如图 4-5 所示。本范例介绍的就是

Gaussian 金字塔。

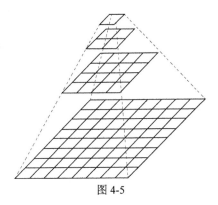

图 4-5

具体代码如下:

```
#include "opencv2/imgproc/imgproc.hpp"
#include "opencv2/highgui/highgui.hpp"
#include <math.h>
#include <stdlib.h>
#include <stdio.h>
using namespace cv;
/// 声明全局变量
Mat src, dst, tmp;
const char* window_name = "Pyramids Demo";
int main(void)
{
   /// 程序使用说明
   printf("\n 图像缩放示范\n ");
   printf("------------------ \n");
   printf(" * [u] -> 放大 \n");
   printf(" * [d] -> 缩小 \n");
   printf(" * [ESC] -> 结束程序 \n \n");
   src = imread("C:\\images\\chicky_512.png");
   if (!src.data)
   {
     printf(" No data! -- Exiting the program \n");
     return -1;
   }
   // 将读取图像放入要处理的变量中
   tmp = src;
   dst = tmp;
   /// 建立窗口
   namedWindow(window_name, WINDOW_AUTOSIZE);
   imshow(window_name, dst);
   /// Loop
   for (;;)
   {
     int c;
     c = waitKey(10);
     if ((char)c == 27)
        break;
     if ((char)c == 'u')
     {
        pyrUp(tmp, dst, Size(tmp.cols * 2, tmp.rows * 2));
```

```
            printf("** 放大：  放大两倍\n");
        } else if ((char)c == 'd') {
            pyrDown(tmp, dst, Size(tmp.cols / 2, tmp.rows / 2));
            printf("** 缩小：  缩小一半\n");
        }
        imshow(window_name, dst);
        // 将结果当成要处理的图
        tmp = dst;
    }
    return 0;
}
```

程序说明

1. pyrUp(InputArray src, OutputArray dst, const Size& dstsize=Size(), int borderType=BORDER_DEFAULT)：向上取样（Upsamples）并模糊化

（1）src：输入图像。

（2）dst：输出图像。

（3）dstsize：输出图像的大小。

（4）borderType：像素外推法边缘的类型。

2. pyrDown(InputArray src, OutputArray dst, const Size& dstsize=Size(), int borderType=BORDER_DEFAULT)：向下取样（Upsamples）并模糊化

（1）src：输入图像。

（2）dst：输出图像。

（3）dstsize：输出图像的大小。

（4）borderType：像素外推法边缘的类型。

执行结果如图 4-6 所示。

图 4-6

读者是否发现放大与缩小都不是无止境的，放大图并不会失真但是没法将图上下移动，学会 HighGUI 就知道 OpenCV 只有基本的窗口功能。而缩小到某种程度就会有残留图像，并且当图最小化时，再进行放大则图已经模糊，无法再查看原图大小。

4.5 基本阈值法

阈值法（Thresholding Operation）是最简单的区隔法，其应用就是相对于某对象将图像区隔进行分析，而区隔是依据对图像像素与背景像素强度差异做区隔。为了区隔相关的像素与其他像素，就使用像素强度值做比较，以适当地区隔重要的像素背景，而此像素强度值就称为阈值。如图 4-7 所示，深灰色区域代表图的像素强度值，而图中粗线就是阈值。

阈值法有下列 5 种。

1. 二进制（binary）阈值法

在执行完二进制的阈值法后，左方与中间的图将会消失，而只有像素强度高于阈值的会保留。保留后的像素强度将设为最大值，结果如图 4-8 所示。

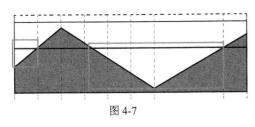

图 4-7

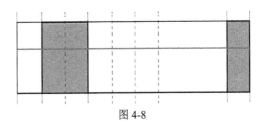

图 4-8

2. 反向（inverted）二进制的阈值法

在执行完反向二进制的阈值法后，只有左方与中间的图会保留，并且只有像素强度低于阈值的才会保留下来。保留后的像素强度将设为最大值，结果如图 4-9 所示。

3. 截断（truncate）阈值法

在执行完截断的阈值法后，大于阈值图的像素强度都改成阈值，结果如图 4-10 所示。

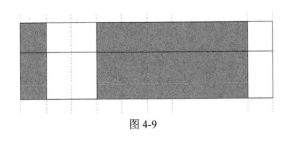

图 4-9

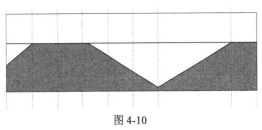

图 4-10

4. 阈值为零法（Threshold to Zero）

在执行阈值为零法之后，只有大于阈值的图会保留，结果如图 4-11 所示。

5. 反向阈值为零法（Threshold to Zero）

在执行完反向阈值为零法后，只有小于阈值的图会保留，结果如图 4-12 所示。

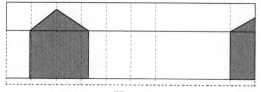

图 4-11

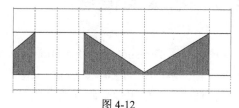

图 4-12

具体代码如下：

```
#include "opencv2/imgproc/imgproc.hpp"
#include "opencv2/highgui/highgui.hpp"
#include <stdlib.h>
#include <stdio.h>
using namespace cv;
/// 声明全局变量
int threshold_value = 0;
int threshold_type = 3;;
int const max_value = 255;
int const max_type = 4;
int const max_BINARY_value = 255;
Mat src, src_gray, dst;
const char* window_name = "Threshold Demo";
const char* trackbar_type = "Type: \n 0: Binary \n 1: Binary Inverted \n 2: Truncate \n 3: To Zero \n 4: To Zero Inverted";
const char* trackbar_value = "Value";
/// 声明函数
void Threshold_Demo(int, void*);
int main(int, char** argv)
{
   /// 加载图形文件
   src = imread("C:\\images\\chicky_512.png", 1);
   /// 将图形文件变换成灰度
   cvtColor(src, src_gray, COLOR_RGB2GRAY);
   /// 建立窗口显示结果
   namedWindow(window_name, WINDOW_AUTOSIZE);
   /// 建立滑杆以选择阈值法类型
   createTrackbar(trackbar_type, window_name, &threshold_type,
     max_type, Threshold_Demo);
   /// 建立滑杆以选择阈值
   createTrackbar(trackbar_value, window_name, &threshold_value,
     max_value, Threshold_Demo);
   /// 主运算开始
   Threshold_Demo(0, 0);
   for (;;)
   {
      int c;
```

```
        c = waitKey(20);
        if ((char)c == 27)
        {
            break;
        }
    }
}
void Threshold_Demo(int, void*)
{
    /* 0: 二进制的
       1: 反向二进制
       2: 截断
       3: 阈值为零
       4: 反向阈值为零
    */
    threshold(src_gray, dst, threshold_value,
        max_BINARY_value, threshold_type);
    imshow(window_name, dst);
}
```

程序说明

threshold(InputArray src, OutputArray dst, double thresh, double maxval, int type)：**阈值法**

（1）src：输入图像。

（2）dst：输出图像。

（3）thresh：阈值。

（4）maxval：使用 THRESH_BINARY 与 THRESH_BINARY_INV 两个阈值法类型的最大值。

（5）type：阈值法类型。

执行结果如图 4-13 所示。

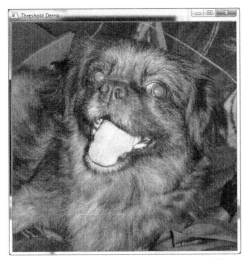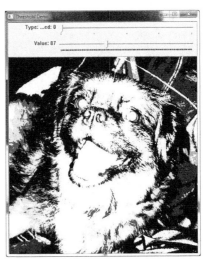

图 4-13

如果程序不将图形文件转成灰度，执行的结果如图 4-14 所示。

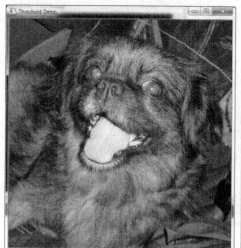

图 4-14

阈值应用

阈值应用的代码如下：

```
#include "opencv2/opencv.hpp"
using namespace cv;
using namespace std;
int main( int argc, char** argv )
{
    namedWindow("orignal", WINDOW_AUTOSIZE);
    namedWindow("threshold", WINDOW_AUTOSIZE);
    // 读取图像
    Mat src = imread("C:\\images\\threshold.png", IMREAD_GRAYSCALE);
    imshow("orignal", src);
    Mat dst;
    // 基本阈值法
    threshold(src, dst, 0, 255, THRESH_BINARY);
    imshow("threshold", dst);
    // Thresholding with maxval set to 128
    threshold(src, dst, 0, 128, THRESH_BINARY);
    imshow("threshold", dst);

    // 阈值设为 127
    threshold(src, dst, 127, 255, THRESH_BINARY);
    imshow("threshold", dst);
    waitKey();

    // 使用 THRESH_BINARY_INV 阈值
    threshold(src, dst, 127, 255, THRESH_BINARY_INV);
    imshow("threshold", dst);

    // 使用 THRESH_TRUNC 阈值
```

```
    threshold(src, dst, 127, 255, THRESH_TRUNC);
    imshow("threshold", dst);
    // 使用 THRESH_TOZERO 阈值
    threshold(src, dst, 127, 255, THRESH_TOZERO);
    imshow("threshold", dst);
    // 使用 THRESH_TOZERO_INV 阈值
    threshold(src, dst, 127, 255, THRESH_TOZERO_INV);
    imshow("threshold", dst);
}
```

原图如图 4-15 所示。

"threshold(src, dst, 0, 255, THRESH_BINARY);"的结果如图 4-16 所示。

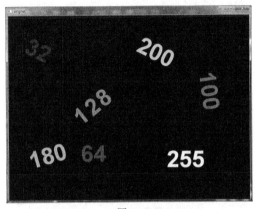

图 4-15　　　　　　　　　　　图 4-16

"threshold(src, dst, 0, 128, THRESH_BINARY);"的结果如图 4-17 所示。

"threshold(src, dst, 127, 255, THRESH_BINARY);"的结果如图 4-18 所示。

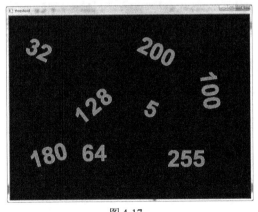 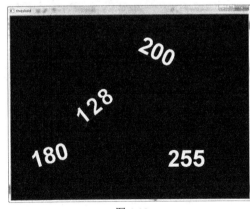

图 4-17　　　　　　　　　　　图 4-18

"threshold(src, dst, 127, 255, THRESH_BINARY_INV);"的结果如图 4-19 所示。

"threshold(src, dst, 127, 255, THRESH_TRUNC);"的结果如图 4-20 所示。

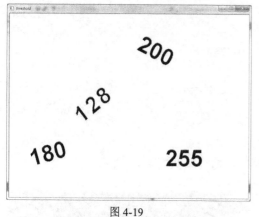

图 4-19

图 4-20

"threshold(src, dst, 127, 255, THRESH_TOZERO);"的结果如图 4-21 所示。
"threshold(src, dst, 127, 255, THRESH_TOZERO_INV);"的结果如图 4-22 所示。

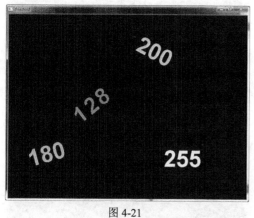

图 4-21

图 4-22

4.6　建立自己的线性滤波器

什么是核（kernel）

基本上核是固定大小，其中心有锚点（anchor point）的数值系数（numerical coefficeints）矩阵，概念如图 4-23 所示。

卷积（convolution）是核与图的每个部分间的运算，其目的是要得知图中某个位置的结果值（resulting value），其运算方式如下：

（1）将核的锚放在已决定的像素上。

（2）像素值乘以核系数再加和。

图 4-23

（3）将结果放在锚点。

（4）在整个图上扫描核并对每个像素重复这些动作。

上面说明的整个运算过程以方程序表示就是：

$$H(x,y)=\sum_{i=0}^{M_{i-1}}\sum_{j=0}^{M_{j-1}}I(x+i-a_i, y+j-a_j)K(i,j)$$

K(i,j)是核系数，而在 OpenCV 提供的函数就是 filter2D，本范例即是介绍 filter2D，具体代码如下。

```cpp
#include "opencv2/imgproc/imgproc.hpp"
#include "opencv2/highgui/highgui.hpp"
#include <stdlib.h>
#include <stdio.h>
using namespace cv;
int main(int, char** argv)
{
  Mat src, dst, kernel;
  Point anchor;
  double delta;
  int ddepth;
  int kernel_size;
  const char* window_name = "filter2D Demo";
  int c;
  src = imread("C:\\images\\lena.jpg");
  if (!src.data)
      return -1;
  namedWindow(window_name, WINDOW_AUTOSIZE);
  /// filter2D 函数的初始参数(argument)
  /// 核
  anchor = Point(-1, -1);
  /// 卷积过程加到每个像素的值
  delta = 0;
  /// dst 的深度，-1 表示与原图相同
  ddepth = -1;
  /// Loop - 每 0.5 秒(500 毫秒)用不同的核大小滤波 (filter) 图像
  int ind = 0;
  for (;;)
  {
    c = waitKey(500);
    /// 按下 'ESC' 结束程序
    if ((char)c == 27)
    {
      break;
    }
    /// 以归一化块滤波器(normalized box filter)来更新核大小
    kernel_size = 3 + 2 * (ind % 5);
    kernel = Mat::ones(kernel_size, kernel_size, CV_32F)
        / (float)(kernel_size*kernel_size);
    /// 开始滤波
    filter2D(src, dst, ddepth, kernel, anchor, delta, BORDER_DEFAULT);
    imshow(window_name, dst);
```

```
        ind++;
    }
    return 0;
}
```

程序说明

filter2D(InputArray src, OutputArray dst, int ddepth, InputArray kernel, Point anchor=Point(-1,-1), double delta=0, int borderType)：用核作卷积(convolves)处理

（1）src：输入图像。

（2）dst：输出图像。

（3）ddepth：输入图像深度（景深），如果是负值（-1），将会与输入图像同深度。

将此值与输入图像深度组合，如表 4-1 所示。

表 4-1

输入图像深度	深度值(ddepth)
CV_8U	-1、CV_16S、CV_32F、CV_64F
CV_16U、CV_16S	-1、CV_32F、CV_64F
CV_32F	-1、CV_32F、CV_64F
CV_64F	-1、CV_64F

（4）kernel：卷积处理核值。

（5）anchor：核的锚点。

（6）delta：产生输出图像前，增加到每个像素的值。

（7）borderType：像素外推法边缘的类型。

执行结果就是图被模糊的过程，如图 4-24 所示。

图 4-24

4.7 将图像加上边框

将图像加上边框的代码如下:

```cpp
#include "opencv2/imgproc/imgproc.hpp"
#include "opencv2/highgui/highgui.hpp"
#include <stdlib.h>
#include <stdio.h>
using namespace cv;
Mat src, dst;
int top, bottom, left, right;
int borderType;
const char* window_name = "copyMakeBorder Demo";
RNG rng(12345);
int main(int, char** argv)
{
   int c;
   src = imread("C:\\images\\lena.jpg");
   if (!src.data)
     return -1;
   imshow("org", src);
   /// 程序使用说明
   printf("\n \t copyMakeBorder Demo: \n");
   printf("\t -------------------- \n");
   printf(" ** 按下 'c' 以随机设置边框 \n");
   printf(" ** 按下 'r' 取消边框 \n");
   printf(" ** 按下 'ESC' 结束程序 \n");
   /// 建立窗口
   namedWindow(window_name, WINDOW_AUTOSIZE);
   /// 边框厚度
   top = (int)(0.05*src.rows);    /// 上
   bottom = (int)(0.05*src.rows); /// 下
   left = (int)(0.05*src.cols);   /// 左
   right = (int)(0.05*src.cols);  /// 右
   dst = src;
   imshow(window_name, dst);
   for (;;)
   {
     c = waitKey(500);
     if ((char)c == 27) {
        break;
     } else if ((char)c == 'c') {
        // 补固定值
        borderType = BORDER_CONSTANT;
     } else if ((char)c == 'r') {
        // 原图值加框
        borderType = BORDER_REPLICATE;
     }
     /// 边框值,使用 rng 随机产生 0 与 254 间的值,不含 255
     Scalar value(rng.uniform(0, 255), rng.uniform(0, 255),
        rng.uniform(0, 255));
     /// value 值只有 borderType = BORDER_CONSTANT 才有用
```

```
        copyMakeBorder(src, dst, top, bottom, left, right,
            borderType, value);
        imshow(window_name, dst);
    }
    return 0;
}
```

执行结果如图 4-25 所示。

（a）执行代码

（b）原图　　　　　　　　　（c）加边框　　　　　　　　（d）取消边框

图 4-25

读者是否注意到边框是在原图外围加上的，当边框去除后是使用原图的值补上边框，所以会以直线显示。

4.8　Sobel 算子

上两节介绍的都是对图进行卷积运算的应用，而计算图的算子（derivatives）是卷积运算中最重要的。其应用在边的检测，也就是图像的像素值有大改变的地方，如图 4-26 所示。

通过 Sobel 函数将上图检测到的部分放大成如图 4-27 所示。

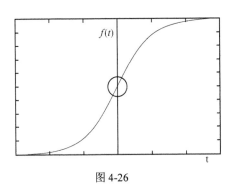

图 4-26

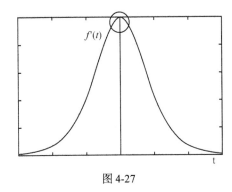

图 4-27

Sobel 运算是一种离散的微分（differentiation）运算，用来计算图像素强度的改变，也就是图的算子。Sobel 运算也结合了高斯（Gaussian）的平滑与微分运算。

具体代码如下：

```cpp
#include "opencv2/imgproc/imgproc.hpp"
#include "opencv2/highgui/highgui.hpp"
#include <stdlib.h>
#include <stdio.h>
using namespace cv;
int main(int, char** argv)
{
    Mat src, src_gray, grad;
    const char* window_name = "Sobel Demo - Simple Edge Detector";
    // Sobel 函数用到的参数
    int scale = 1;
    int delta = 0;
    int ddepth = CV_16S;
    /// 加载图形文件
    src = imread("C:\\images\\lena.jpg");
    if (!src.data)
        return -1;
    // 对原图使用 Gaussian 模糊法，使用 Kernel Size(3, 3)先去除噪声
    GaussianBlur(src, src, Size(3, 3), 0, 0, BORDER_DEFAULT);
    /// 变换成灰度
    cvtColor(src, src_gray, COLOR_RGB2GRAY);
    /// 建立窗口
    namedWindow(window_name, WINDOW_AUTOSIZE);
    /// 建立 grad_x 与 grad_y
    Mat grad_x, grad_y;
    Mat abs_grad_x, abs_grad_y;
    /// 倾斜度 (Gradient) X
    /*
    Scharr( src_gray, grad_x, ddepth, 1, 0,
        scale, delta, BORDER_DEFAULT );
    */
    Sobel(src_gray, grad_x, ddepth, 1, 0, 3,
        scale, delta, BORDER_DEFAULT);

    /// 变换成绝对值 8 位图像
    convertScaleAbs(grad_x, abs_grad_x);
    /// 倾斜度 Y
```

```
    /*
    Scharr( src_gray, grad_y, ddepth, 0, 1,
       scale, delta, BORDER_DEFAULT );
    */
    Sobel(src_gray, grad_y, ddepth, 0, 1, 3,
       scale, delta, BORDER_DEFAULT);

    convertScaleAbs(grad_y, abs_grad_y);
    /// 总倾斜度（大约的）
    addWeighted(abs_grad_x, 0.5, abs_grad_y, 0.5, 0, grad);
    imshow(window_name, grad);
    waitKey(0);
    return 0;
}
```

程序说明

1. Scharr(InputArray src, OutputArray dst, int ddepth, int dx, int dy, double scale=1, double delta=0, int borderType=BORDER_DEFAULT)：计算图像 X 或 Y 轴算子值

　　（1）src：输入图像。

　　（2）dst：输出图像。

　　（3）ddepth：输出图像深度。

　　（4）dx：X 轴算子。

　　（5）dy：Y 轴算子。

　　（6）scale：计算算子值的缩放因子（scale factor），此参数为可选项。

　　（7）delta：加到输出图像的差异值。

　　（8）borderType：边缘形态。

2. Sobel(InputArray src, OutputArray dst, int ddepth, int dx, int dy, int ksize=3, double scale=1, double delta=0, int borderType=BORDER_DEFAULT)：计算第一、第二、第三、或混合的图像算子

　　（1）src：输入图像。

　　（2）dst：输出图像。

　　（3）ddepth：输出图像深度。

　　（4）dx：X 轴算子。

　　（5）dy：Y 轴算子。

　　（6）ksize：Sobel 核大小，值可以是 1、3、5 或 7。

　　（7）scale：计算算子值的缩放因子（scale factor）；此参数为可选项。

　　（8）delta：加到输出图像的差异值。

　　（9）borderType：边缘形态。

3. convertScaleAbs(InputArray src, OutputArray dst, double alpha=1, double beta=0)：计算绝对值并将结果变换成 8 位

（1）src：输入图像。
（2）dst：输出图像。
（3）alpha：缩放因子（scale factor），该参数为可选项。
（4）beta：加到缩放值的差异值，该参数为可选项。

该函数依次对图执行缩放，使用绝对值与变换成 8 位无正负号的值（其实在第二个绝对值操作的时候已去除正负号）。

执行结果如图 4-28 所示。

如果将渐层函数：

```
/// 渐层 X
Sobel( src_gray, grad_x, ddepth, 1, 0,
   3, scale, delta, BORDER_DEFAULT );
```

与

```
/// 渐层 Y
Sobel( src_gray, grad_y, ddepth, 0, 1,
   3, scale, delta, BORDER_DEFAULT );
```

改成：

```
/// 渐层 X
Scharr( src_gray, grad_x, ddepth, 1, 0,
   scale, delta, BORDER_DEFAULT );
```

与

```
/// 渐层 Y
Scharr( src_gray, grad_y, ddepth, 0, 1,
   scale, delta, BORDER_DEFAULT );
```

结果如图 4-29 所示。

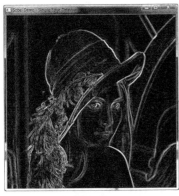
图 4-28

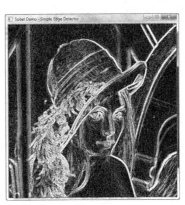
图 4-29

再改成:

```
int kernel_size = 3;
/// 渐层 X
Laplacian(src_gray, grad_x, ddepth, kernel_size,
   scale, delta, BORDER_DEFAULT);
```

与

```
/// 渐层 Y
Laplacian(src_gray, grad_y, ddepth, kernel_size,
   scale, delta, BORDER_DEFAULT);
```

结果如图 4-30 所示。

将所有渐层部分改成:

```
Canny( src_gray, grad, 50, 200, 3 );
```

结果如图 4-31 所示。

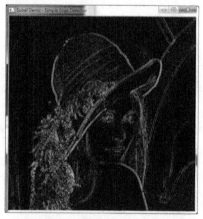

图 4-30

图 4-31

4.9 拉普拉斯运算

拉普拉斯运算的代码如下:

```
#include "opencv2/imgproc/imgproc.hpp"
#include "opencv2/highgui/highgui.hpp"
#include <stdlib.h>
#include <stdio.h>
using namespace cv;
int main(int, char** argv)
{
 Mat src, src_gray, dst;
   /// Laplacian 函数用到的参数
   int kernel_size = 3;
   int scale = 1;
```

```cpp
    int delta = 0;
    int ddepth = CV_16S;
    const char* window_name = "Laplace Demo";
    src = imread("C:\\images\\lena.jpg");
    if (!src.data)
        return -1;
    /// 对原图使用Gaussian模糊法，使用Kernel Size(3, 3)先去除噪声
    GaussianBlur(src, src, Size(3, 3), 0, 0, BORDER_DEFAULT);
    /// 转成灰度
    cvtColor(src, src_gray, COLOR_RGB2GRAY);
    /// 新建窗口
    namedWindow(window_name, WINDOW_AUTOSIZE);
    /// 使用 Laplace 函数
    Mat abs_dst;
    Laplacian(src_gray, dst, ddepth, kernel_size,
        scale, delta, BORDER_DEFAULT);
    // 拉普拉斯函数处理后转成绝对值
    convertScaleAbs(dst, abs_dst);
    imshow(window_name, abs_dst);
    waitKey(0);
    return 0;
}
```

程序说明

Laplacian(InputArray src, OutputArray dst, int ddepth, int ksize=1, double scale=1, double delta=0, int borderType=BORDER_DEFAULT)：对图像进行拉普拉斯运算

（1）src：输入图像。

（2）dst：输出图像。

（3）ddepth：输出图像深度。

（4）ksize：计算第二算子（second-derivate）光圈大小（aperture size）。

（5）scale：计算拉普拉斯值的缩放因子。

（6）delta：加到输出图像的差异值。

（7）borderType：边缘类形。

执行结果如图 4-32 所示。

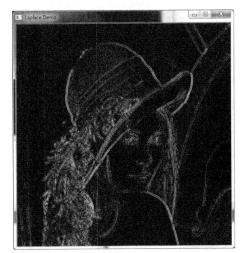

图 4-32

4.10 Canny 图像边缘检测

这是由 John F. Canny 开发的，因而以其名称来命名。这个方法满足了下列 3 种条件。

（1）低错误率：仅在存在的边缘上，是不错的检测。

（2）局部化良好：检测到像素边缘之间的距离最小，实际边缘的像素也最小。

（3）简单反应：每个边缘只有一个检测结果。

检测步骤如下：
（1）用 Gaussian 滤波器去除噪声。
（2）用类似 Sobel 方法寻找图像倾斜度的强度。
（3）采用非极大值抑制（non-maximum suppression），这个步骤会删除认为不是边缘的像素，所以只留下细线。
（4）滞后作用：这是最后的步骤中使用的两种临界法。
- 如果像素倾斜度高于上阈值就被认为是边缘。
- 如果像素倾斜度低于下阈值就被舍去。
- 如果像素倾斜度介于两个阈值之间，而相连的像素强度高于上阈值就被认为是边缘。

该方法的发明者认为上下阈值分别是 2:1 与 3:1。
具体代码如下：

```cpp
#include "opencv2/imgproc/imgproc.hpp"
#include "opencv2/highgui/highgui.hpp"
#include <stdlib.h>
#include <stdio.h>
using namespace cv;
/// 声明全局变量
Mat src, src_gray;
Mat dst, detected_edges;
int edgeThresh = 1;
int lowThreshold;
int const max_lowThreshold = 100;
int ratio = 3;
int kernel_size = 3;
const char* window_name = "Edge Map";
// 滑杆回调函数 - Canny 以 1:3 作为掩码输入
// 因为用不到传入的参数，所以没有命名参数变量名
static void CannyThreshold(int, void*)
{
    /// 以 3x3 的核减少噪声
    blur(src_gray, detected_edges, Size(3, 3));
    /// Canny 检测器
    Canny(detected_edges, detected_edges, lowThreshold,
        lowThreshold*ratio, kernel_size);
    /// 内容全为 0
    dst = Scalar::all(0);
    /// 将 src 以 detected_edges 为掩码复制到 dst
    src.copyTo(dst, detected_edges);
    /// 显示掩码后结果
    imshow(window_name, dst);
}
int main(int, char** argv)
{
    src = imread("C:\\images\\lena.jpg");
    if (!src.data)
        return -1;
    /// 建立一个与原图大小与形态一样的矩阵
    dst.create(src.size(), src.type());
    /// 转成灰度
```

```
    cvtColor(src, src_gray, COLOR_BGR2GRAY);
    /// 建立窗口
    namedWindow(window_name, WINDOW_AUTOSIZE);
    /// 建立滑杆用来作为掩码值得输入
    createTrackbar("Min Threshold:", window_name, &lowThreshold,
       max_lowThreshold, CannyThreshold);
    /// 掩码后显示结果
    CannyThreshold(0, 0);
    waitKey(0);
    return 0;
}
```

程序说明

Canny(InputArray image, OutputArray edges, double threshold1, double threshold2, int apertureSize=3, bool L2gradient=false)：用 Canny 算法寻找图像边缘

（1）image：输入图像。

（2）edges：输出的边缘图。

（3）threshold1：滞后程序的第一个阈值。

（4）threshold2：滞后程序的第二个阈值。

（5）apertureSize：Sobel 运算的光圈大小。

（6）L2gradient：是否更精准的标志。

执行结果如图 4-33 所示。

 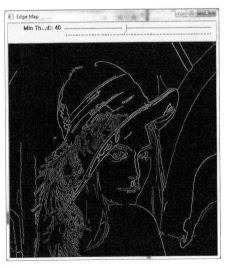

图 4-33

按照程序看应该是灰度图像，将程序改为直接显示检测边缘结果的图像，即 imshow(window_name, detected_edges)，也就是省略 src.copyTo。

执行结果如图 4-34 所示。

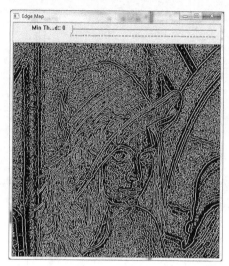
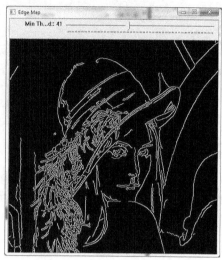

图 4-34

4.11 霍夫线变换

霍夫线变换（Hough Line Transform）是直线检测，要使用此变换时，需要先进行边缘检测的前置处理。此变换是使用极坐标系统（Polar Coordinate System）来表达直线，而数学方程式就是：

$$r = x \cos \theta + y \sin \theta$$

若以坐标图说明，则是要表达右斜线，可以使用与右斜线呈直角的左斜线的两个参数 (r,θ)。右斜线每一点的数学方程式即为 $r_\theta = x_0 \cdot \cos \theta + y_0 \cdot \sin \theta$，如图 4-35 所示。

将图 4-35 中笛卡儿坐标图以此方程式表达成极坐标图就是正弦曲线，但是必须考虑条件为 $r>0$ 与 $0<\theta<2\pi$，如图 4-36 所示。

将此坐标变换用于图像的每个点，如果线段的两个不同的点在极坐标内有交叉，就表示这两点属于同一条线。

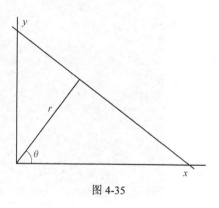

图 4-35

概括起来，就是线可以由不同曲线间的交叉点数来检测，曲线交叉越多就表示线的点越多，以最小交叉数当作阈值来检测线，这就是霍夫线变换。当交叉数高于阈值，则表示有线存在。

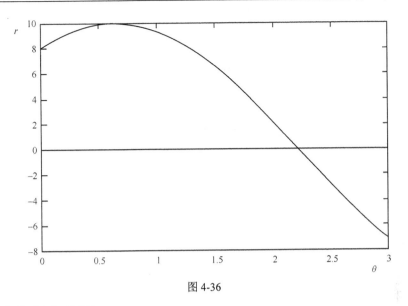

图 4-36

霍夫线变换有以下两种。

（1）标准的霍夫变换。

其内容如上所述，在 OpenCV 就是 HoughLines 函数。

（2）概率（Probabilistic）霍夫变换。

这是更有效的变换。而在 OpenCV 就是 HoughLinesP 函数。

霍夫线变换的代码如下：

```cpp
#include "opencv2/highgui/highgui.hpp"
#include "opencv2/imgproc/imgproc.hpp"
#include <iostream>
#include <stdio.h>
using namespace cv;
using namespace std;
/// 声明全局变量
Mat src, edges;
Mat src_gray;
Mat standard_hough, probabilistic_hough;
int min_threshold = 50;
int max_trackbar = 150;
const char* standard_name = "Standard Hough Lines Demo";
const char* probabilistic_name = "Probabilistic Hough Lines Demo";
int s_trackbar = max_trackbar;
int p_trackbar = max_trackbar;
/// 声明函数
void Standard_Hough(int, void*);
void Probabilistic_Hough(int, void*);
int main(int, char** argv)
{
    src = imread("C:\\images\\building.jpg", 1);
    if (src.empty())
        return -1;
    /// 转成灰度
```

```cpp
    cvtColor(src, src_gray, COLOR_RGB2GRAY);
    /// 用 Canny 进行图像边缘检测
    Canny(src_gray, edges, 50, 200, 3);
    /// 用滑杆作为掩码输入
    char thresh_label[50];
    sprintf(thresh_label, "Thres: %d + input", min_threshold);
    namedWindow(standard_name, WINDOW_AUTOSIZE);
    createTrackbar(thresh_label, standard_name,
        &s_trackbar, max_trackbar, Standard_Hough);
    namedWindow(probabilistic_name, WINDOW_AUTOSIZE);
    createTrackbar(thresh_label, probabilistic_name,
        &p_trackbar, max_trackbar, Probabilistic_Hough);
    /// 开始操作
    Standard_Hough(0, 0);
    Probabilistic_Hough(0, 0);
    waitKey(0);
    return 0;
}
/// 标准的 Hough 变换
void Standard_Hough(int, void*)
{
    vector<Vec2f> s_lines;
    cvtColor(edges, standard_hough, CV_GRAY2BGR);
    HoughLines(edges, s_lines, 1, CV_PI / 180,
        min_threshold + s_trackbar, 0, 0);
    /// 1. 用标准的 Hough 变换
    for (size_t i = 0; i < s_lines.size(); i++)
    {
        float r = s_lines[i][0], t = s_lines[i][1];
        double cos_t = cos(t), sin_t = sin(t);
        double x0 = r*cos_t, y0 = r*sin_t;
        double alpha = 1000;
        Point pt1(cvRound(x0 + alpha*(-sin_t)),
            cvRound(y0 + alpha*cos_t));
        Point pt2(cvRound(x0 - alpha*(-sin_t)),
            cvRound(y0 - alpha*cos_t));
        line(standard_hough, pt1, pt2, Scalar(255, 0, 0), 3, CV_AA);
    }
    /// 显示结果
    imshow(standard_name, standard_hough);
}
/// 概率 Hough 变换
void Probabilistic_Hough(int, void*)
{
    vector<Vec4i> p_lines;
    cvtColor(edges, probabilistic_hough, CV_GRAY2BGR);
    HoughLinesP(edges, p_lines, 1, CV_PI / 180,
        min_threshold + p_trackbar, 30, 10);
    /// 2. 用概率的 Hough 变换
    for (size_t i = 0; i < p_lines.size(); i++)
    {
        Vec4i l = p_lines[i];
        line(probabilistic_hough, Point(l[0], l[1]),
            Point(l[2], l[3]), Scalar(255, 0, 0), 3, CV_AA);
    }
    /// 显示结果
    imshow(probabilistic_name, probabilistic_hough);
}
```

程序说明

1. HoughLines(InputArray image, OutputArray lines, double rho, double theta, int threshold, double srn=0, double stn=0)：用霍夫变换在二元图像中寻找线

（1）image：输入图像，此图像可能被此函数改变。

（2）lines：输出线的向量值（ρ,θ）；ρ 是与坐标原点（0,0）的距离（坐标原点就是图的左上角），θ 是旋转角度（0 是垂直线，$\pi/2$ 是水平线）。

（3）rho：像素中距离解析（distance resolution）的累加器（accumulator）。

（4）theta：累加器的角解析（angle resolution）弧度。

（5）threshold：累加器的阈值参数（threshold parameter）。

（6）srn：用于多级别（multi-scale）霍夫变换，是距离解析 rho 的除数（divisor）。

（7）stn：用于多级别（multi-scale）霍夫变换，是角解析 theta 的除数。

2. HoughLinesP(InputArray image, OutputArray lines, double rho, double theta, int threshold, double minLineLength=0, double maxLineGap=0)：用概率的霍夫变换在二元图像中寻找线

（1）image：输入图像，此图像可能被此函数改变。

（2）lines：输出线的向量值（ρ,θ）；ρ 是与坐标原点（0,0）的距离（坐标原点就是图的左上角），θ 是旋转角度（0 是垂直线，$\pi/2$ 是水平线）。

（3）rho：像素中距离解析（distance resolution）的累加器（accumulator）。

（4）theta：累加器的角解析（angle resolution）弧度。

（5）threshold：累加器的阈值参数（threshold parameter）。

（6）minLineLength：线的最小长度。

（7）maxLineGap：线中两点间允许的最大间格（gap）。

执行结果如图 4-37 所示。

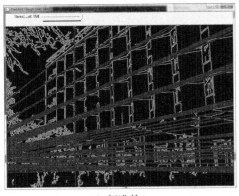 

（a）标准的　　　　　　　　　　（b）概率的

图 4-37

实际应用程序如下，效果如图 4-38 所示。

```cpp
#include "opencv2/highgui/highgui.hpp"
#include "opencv2/imgproc/imgproc.hpp"
#include <iostream>
using namespace cv;
using namespace std;
int main(int argc, char** argv)
{
    const char* filename = argc >= 2 ? argv[1] : "c:/images/pic1.png";
    Mat src = imread(filename, 0);
    if(src.empty())
    {
        help();
        cout << "can not open " << filename << endl;
        return -1;
    }
    Mat dst, cdst;
    Canny(src, dst, 50, 200, 3);
    cvtColor(dst, cdst, COLOR_GRAY2BGR);
#if 0
    vector<Vec2f> lines;
    HoughLines(dst, lines, 1, CV_PI/180, 100, 0, 0 );
    for( size_t i = 0; i < lines.size(); i++ )
    {
        float rho = lines[i][0], theta = lines[i][1];
        Point pt1, pt2;
        double a = cos(theta), b = sin(theta);
        double x0 = a*rho, y0 = b*rho;
        pt1.x = cvRound(x0 + 1000*(-b));
        pt1.y = cvRound(y0 + 1000*(a));
        pt2.x = cvRound(x0 - 1000*(-b));
        pt2.y = cvRound(y0 - 1000*(a));
        line( cdst, pt1, pt2, Scalar(0,0,255), 3, CV_AA);
    }
#else
    vector<Vec4i> lines;
    HoughLinesP(dst, lines, 1, CV_PI/180, 50, 50, 10 );
    for( size_t i = 0; i < lines.size(); i++ )
    {
        Vec4i l = lines[i];
        line( cdst, Point(l[0], l[1]), Point(l[2], l[3]),
            Scalar(0,0,255), 3, CV_AA);
    }
#endif
    imshow("source", src);
    imshow("detected lines", cdst);
    waitKey();
    return 0;
}
```

（a）原图　　　　　　　　　　　（b）检测结果

图 4-38

4.12　霍夫圆变换

霍夫圆的代码如下：

```cpp
#include "opencv2/highgui/highgui.hpp"
#include "opencv2/imgproc/imgproc.hpp"
#include <iostream>
using namespace cv;
namespace
{
    // 窗口与滑杆名称
    const std::string windowName
        = "Hough Circle Detection Demo";
    const std::string cannyThresholdTrackbarName
        = "Canny threshold";
    const std::string accumulatorThresholdTrackbarName
        = "Accumulator Threshold";
    // 起始值与最大值
    const int cannyThresholdInitialValue = 200;
    const int accumulatorThresholdInitialValue = 50;
    const int maxAccumulatorThreshold = 200;
    const int maxCannyThreshold = 255;
    void HoughDetection(const Mat& src_gray,
            const Mat& src_display, int cannyThreshold,
            int accumulatorThreshold)
    {
        std::vector<Vec3f> circles;
        // 执行检测，circle 为检测结果的值
        HoughCircles(src_gray, circles, CV_HOUGH_GRADIENT,
            1, src_gray.rows / 8, cannyThreshold,
            accumulatorThreshold, 0, 0);
        // 绘制检测到的圆
        Mat display = src_display.clone();
        for (size_t i = 0; i < circles.size(); i++)
        {
            Point center(cvRound(circles[i][0]), cvRound(circles[i][1]));
            int radius = cvRound(circles[i][2]);
```

```cpp
            // 圆的中心
            circle(display, center, 3, Scalar(0, 255, 0), -1, 8, 0);
            // 圆的轮廓
            circle(display, center, radius, Scalar(0, 0, 255), 3, 8, 0);
        }
        // 显示结果
        imshow(windowName, display);
    }
}
int main(int argc, char** argv)
{
    Mat src, src_gray;
    src = imread("C:\\images\\opencv-logo.png", 1);
    if (!src.data)
    {
        std::cerr << "文档读取失败\n";
        return -1;
    }
    // 转成灰度
    cvtColor(src, src_gray, COLOR_BGR2GRAY);
    // 减少噪声
    GaussianBlur(src_gray, src_gray, Size(9, 9), 2, 2);
    // 声明参数改变前的初始值
    int cannyThreshold = cannyThresholdInitialValue;
    int accumulatorThreshold = accumulatorThresholdInitialValue;
    // 建立窗口
    namedWindow(windowName, WINDOW_AUTOSIZE);
    // 建立滑杆
    createTrackbar(cannyThresholdTrackbarName, windowName,
        &cannyThreshold, maxCannyThreshold);
    createTrackbar(accumulatorThresholdTrackbarName,
        windowName, &accumulatorThreshold,
        maxAccumulatorThreshold);
    // 将执行结果显示
    int key = 0;
    while (key != 'q' && key != 'Q')
    {
        // 检查参数不可为 0
        cannyThreshold = std::max(cannyThreshold, 1);
        accumulatorThreshold = std::max(accumulatorThreshold, 1);
        // 执行检测并更新显示结果
        HoughDetection(src_gray, src, cannyThreshold, accumulatorThreshold);
        key = waitKey(10);
    }
    return 0;
}
```

程序说明

HoughCircles(InputArray image, OutputArray circles, int method, double dp, double minDist, double param1=100, double param2=100, int minRadius=0, int maxRadius=0)：用霍夫变换在二元图像中寻找圆

（1）image：输入的灰度图像。

（2）circles：找到圆的输出向量（x,y,r），x 与 y 是坐标值，r 是半径。

（3）method：检测方法，目前只有一种方法 CV_HOUGH_GRADIENT。

（4）dp：累加器解析对图像解析的反向比（inverse ratio）。

- dp = 1：累加器与原图有相同的解析。
- dp = 2：累加器是原图一半的宽与高。

（5）minDist：检测到的圆圆心间最小距离，如果值太小，圆可能检测错误；如果值太大，圆可能被忽略。

（6）param1：指定方法的第一个参数，如果是 CV_HOUGH_GRADIENT 就是传给 canny 函数两个阈值中较高的阈值。

（7）param2：指定方法的第二个参数，如果是 CV_HOUGH_GRADIENT 就是圆心累加器阈值，越小越会检测到错误的圆。

（8）minRadius：圆的最小半径。

（9）maxRadius：圆的最大半径。

执行结果如图 4-39 所示。

实际应用程序如下，结果如图 4-40 所示。

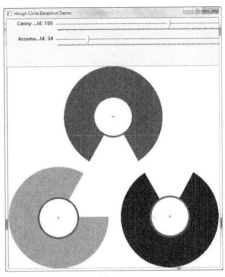

图 4-39

（a）原图　　　　　　　　　（b）变换后

图 4-40

```
#include "opencv2/highgui/highgui.hpp"
#include "opencv2/imgproc/imgproc.hpp"
#include <iostream>
using namespace cv;
using namespace std;
int main(int argc, char** argv)
{
    const char* filename = argc >= 2 ?
        argv[1] : "c:/images/board.jpg";
```

```
    Mat img = imread(filename, 0);
    if(img.empty())
    {
        cout << "can not open " << filename << endl;
        return -1;
    }
    Mat cimg;
    medianBlur(img, img, 5);
    cvtColor(img, cimg, COLOR_GRAY2BGR);
    vector<Vec3f> circles;
    HoughCircles(img, circles, CV_HOUGH_GRADIENT, 1, 10,
                 100, 30, 1, 30);
    for( size_t i = 0; i < circles.size(); i++ )
    {
        Vec3i c = circles[i];
        circle( cimg, Point(c[0], c[1]),
        c[2], Scalar(0,0,255), 3, CV_AA);
        circle( cimg, Point(c[0], c[1]),
        2, Scalar(0,255,0), 3, CV_AA);
    }
    imshow("detected circles", cimg);
    waitKey();
    return 0;
}
```

4.13 重映射

所谓重映射（remapping）是取出图像中某位置的像素，放到新图像的另一位置。

重映射的代码如下：

```
#include "opencv2/highgui/highgui.hpp"
#include "opencv2/imgproc/imgproc.hpp"
#include <iostream>
#include <stdio.h>
using namespace cv;
/// 声明全局变量
Mat src, dst;
Mat map_x, map_y;
const char* remap_window = "Remap demo";
int ind = 0;
/// 声明函数
void update_map(void);
int main(int, char** argv)
{
    src = imread("C:\\images\\lena.jpg", 1);
    /// 用原图的大小建立 dst, map_x and map_y
    dst.create(src.size(), src.type());
    map_x.create(src.size(), CV_32FC1);
    map_y.create(src.size(), CV_32FC1);
    /// 建立窗口
    namedWindow(remap_window, WINDOW_AUTOSIZE);
    for (;;)
```

```cpp
    {
        /// 等 1 秒
        int c = waitKey(1000);
        // 按键是否为 ESC
        if ((char)c == 27)
            break;
        /// 更新 map_x & map_y
        update_map();
        remap(src, dst, map_x, map_y, CV_INTER_LINEAR,
            BORDER_CONSTANT, Scalar(0, 0, 0));
        imshow(remap_window, dst);
    }
    return 0;
}
void update_map(void)
{
    ind = ind % 4;
    for (int j = 0; j < src.rows; j++)
    {
        for (int i = 0; i < src.cols; i++)
        {
            switch (ind)
            {
                case 0:
                    if (i > src.cols*0.25 && i < src.cols*0.75
                        && j > src.rows*0.25 && j < src.rows*0.75)
                    {
                        map_x.at<float>(j, i) = 2 * (i - src.cols*0.25f) + 0.5f;
                        map_y.at<float>(j, i) = 2 * (j - src.rows*0.25f) + 0.5f;
                    }
                    else
                    {
                        map_x.at<float>(j, i) = 0;
                        map_y.at<float>(j, i) = 0;
                    }
                    break;
                case 1:
                    map_x.at<float>(j, i) = (float)i;
                    map_y.at<float>(j, i) = (float)(src.rows - j);
                    break;
                case 2:
                    map_x.at<float>(j, i) = (float)(src.cols - i);
                    map_y.at<float>(j, i) = (float)j;
                    break;
                case 3:
                    map_x.at<float>(j, i) = (float)(src.cols - i);
                    map_y.at<float>(j, i) = (float)(src.rows - j);
                    break;
            }
        }
    }
    ind++;
}
```

程序说明

remap(InputArray src, OutputArray dst, InputArray map1, InputArray map2, int interpolation, int borderMode=BORDER_CONSTANT, const Scalar& borderValue=Scalar())：对图像进行一般性的几何变换

（1）src：原图像。

（2）dst：变换结果图像，与 map1 同大小，与 src 同类型。

（3）map1：在 (x, y) 点的第一个图像。

（4）map2：第二个图像。

（5）interpolation：插值法。

INTER_NEAREST：最近插值法。

INTER_LINEAR：双线性（bilinear）插值法，这是默认值。

INTER_AREA：基于局部像素（pixel area relation）重采样。

INTER_CUBIC：在 4×4 像素范围区双三次（bicubic）插值法。

INTER_LANCZOS4：在 8×8 像素范围区 Lanczos 插值法。

（6）borderMode：像素外推法（extrapolation method）边缘的模式，如果值是 BORDER_RANSPARENT，图的外缘不会改变。

（7）borderValue：固定边缘的值，默认值是 0。

程序显示的 4 种结果如图 4-41 所示。

（a）ind = 0，大小减半

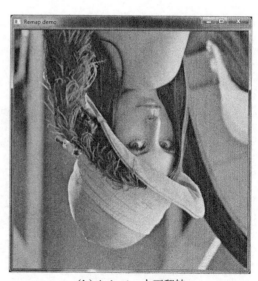

（b）ind = 1，上下翻转

图 4-41

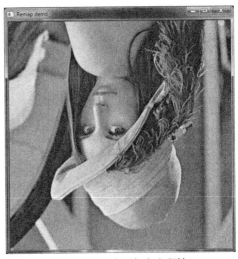

（c）ind = 2，左右翻转　　　　　　　（d）ind = 3，上下加左右翻转

图 4-41（续）

4.14　仿射变换

仿射变换是矩阵相乘（线性变换）再与向量相加（位移）。所以仿射变换的应用有旋转（线性变换）、平移（向量相加）和缩放（线性变换）。

仿射变换的代码如下：

```
#include "opencv2/highgui/highgui.hpp"
#include "opencv2/imgproc/imgproc.hpp"
#include <iostream>
#include <stdio.h>
using namespace cv;
using namespace std;
/// 声明全局变量
const char* source_window = "Source image";
const char* warp_window = "Warp";
const char* warp_rotate_window = "Warp + Rotate";
int main(int, char** argv)
{
   Point2f srcTri[3];
   Point2f dstTri[3];
   Mat rot_mat(2, 3, CV_32FC1);
   Mat warp_mat(2, 3, CV_32FC1);
   Mat src, warp_dst, warp_rotate_dst;
   /// 加载图形文件
   src = imread("C:\\images\\lena.jpg", 1);
   /// 设置结果与原图相同大小
   warp_dst = Mat::zeros(src.rows, src.cols, src.type());
   /// 设置计算仿射变换的三个点
```

```cpp
        srcTri[0] = Point2f(0, 0);
        srcTri[1] = Point2f(src.cols - 1.f, 0);
        srcTri[2] = Point2f(0, src.rows - 1.f);
        dstTri[0] = Point2f(src.cols*0.0f, src.rows*0.33f);
        dstTri[1] = Point2f(src.cols*0.85f, src.rows*0.25f);
        dstTri[2] = Point2f(src.cols*0.15f, src.rows*0.7f);
        /// 取得仿射变换的结果
        warp_mat = getAffineTransform(srcTri, dstTri);
        /// 将仿射变换的结果应用到原图
        warpAffine(src, warp_dst, warp_mat, warp_dst.size());
        /** 翘曲(warp)后旋转 */
        /// 根据图的中心计算旋转矩阵
        Point center = Point(warp_dst.cols / 2, warp_dst.rows / 2);
        double angle = -50.0;
        double scale = 0.6;
        /// 取得旋转矩阵
        rot_mat = getRotationMatrix2D(center, angle, scale);
        /// 旋转图
        warpAffine(warp_dst, warp_rotate_dst, rot_mat, warp_dst.size());
        /// 显示结果
        namedWindow(source_window, WINDOW_AUTOSIZE);
        imshow(source_window, src);
        namedWindow(warp_window, WINDOW_AUTOSIZE);
        imshow(warp_window, warp_dst);
        namedWindow(warp_rotate_window, WINDOW_AUTOSIZE);
        imshow(warp_rotate_window, warp_rotate_dst);
        /// 等待按键
        waitKey(0);
        return 0;
}
```

程序说明

1. Mat getAffineTransform(InputArray src, InputArray dst)：由三对点（three pairs of the corresponding points）计算仿射变换

或是：

Mat getAffineTransform(const Point2f src[], const Point2f dst[])

（1）src：原图像三角形顶点坐标。

（2）dst：变换结果图像对应的三角形顶点坐标。

2. Mat getRotationMatrix2D(Point2f center, double angle, double scale)：计算 2D 旋转的 Affine 矩阵

（1）center：原图像旋转中心。

（2）angle：旋转角度，正值是反时针旋转（坐标原点在左上角）。

（3）scale：等向缩放因子（isotropic scale factor）。

3. warpAffine(InputArray src, OutputArray dst, InputArray M, Size dsize, int flags=INTER_LINEAR, int borderMode=BORDER_CONSTANT, const Scalar& borderValue= Scalar())：对图像执行仿射变换

（1）src：输入图像。

（2）dst：输出图像。

（3）M：2×3 变换矩阵。

（4）dsize：输出图像大小。

（5）flags：像素外推法的组合，如果是 WARP_INVERSE，则表示 M 是反转变换。

（6）borderMode：像素外推法边缘的模式，如果值是 BORDER_TRANSPARENT，则图的外缘不会改变。

（7）borderValue：固定边缘的值，默认值是 0。

执行结果如图 4-42 所示。

（a）原图　　　　　　　　（b）翘曲　　　　　　　　（c）翘曲加旋转

图 4-42

4.15　直方图分布平等化

图像的直方图分布（Histogram）是用来表示图像素强度的分布，将像素强度以数字量化展现，如图 4-43 所示。

直方图分布平等化（Equalization）是改善图像对比的方法，目的是为了扩增像素强度范围。例如，图 4-43 强度太过集中，经过平等化之后就成为图 4-44。图 4-44 中左右圈起来的部

 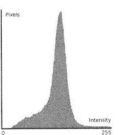

图 4-43

分就是强度分布较少的区域。

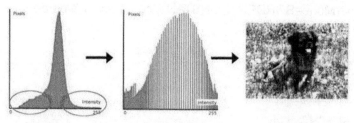

图 4-44

Photoshop 调整直方图是使用手动方式调整,而此处则是用程序来处理的。

```
#include "opencv2/highgui/highgui.hpp"
#include "opencv2/imgproc/imgproc.hpp"
#include <iostream>
#include <stdio.h>
using namespace cv;
using namespace std;
int main(int, char** argv)
{
   Mat src, dst;
   const char* source_window = "Source image";
   const char* equalized_window = "Equalized Image";
   /// 加载图形文件
   src = imread("C:\\images\\lena.jpg", 1);
   if (!src.data)
      return -1;
   /// 变换成灰度
   cvtColor(src, src, COLOR_BGR2GRAY);
   /// 使用直方图分布图 (Histogram)进行平等化(Equalization)
   equalizeHist(src, dst);
   /// 显示结果
   namedWindow(source_window, WINDOW_AUTOSIZE);
   namedWindow(equalized_window, WINDOW_AUTOSIZE);
   imshow(source_window, src);
   imshow(equalized_window, dst);
   /// 等待按键
   waitKey(0);
   return 0;
}
```

程序说明

equalizeHist(InputArray src, OutputArray dst):直方图平等化

(1) src:输入图像。

(2) dst:输出图像。

执行结果如图 4-45 所示。

（a）原图　　　　　　　　　　　　（b）直方图平等化后

图 4-45

4.16 直方图分布计算

直方图分布是收集资料组合成预先定义小类（bins）的次数。而这里所谓的资料是不限于像素强度值，而是可以用来描述图像的任何特征资料，例如倾斜度、方向等。

假设图的像素值如图 4-46 所示。

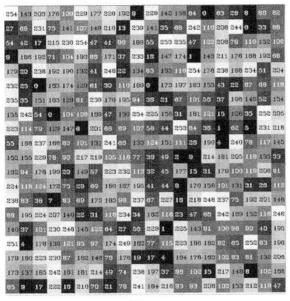

图 4-46

为了有组织地计算各种像素的次数，我们会先将各种数值归类，每个归类的类别就称为小类（bins）。归类后如图 4-47 所示，从 bin1 到 bin16。OpenCV 提供 calcHist 函数进行此计算。

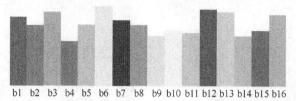

图 4-47

具体代码如下：

```cpp
#include "opencv2/highgui/highgui.hpp"
#include "opencv2/imgproc/imgproc.hpp"
#include <iostream>
#include <stdio.h>
using namespace std;
using namespace cv;
int main(int, char** argv)
{
    Mat src, dst;
    /// 加载图形文件
    src = imread("C:\\images\\lena.jpg", 1);
    if (!src.data)
        return -1;
    /// 分离图像的 3 个颜色通道( B, G and R )
    vector<Mat> bgr_planes;
    split(src, bgr_planes);
    /// 设置 bins，由 0 到 255
    int histSize = 256;
    /// 设置颜色范围值 ( B,G,R )
    float range[] = { 0, 256 };
    const float* histRange = { range };
    /// 设置类别大小一致(uniform)，一开始先清除直方图
    bool uniform = true; bool accumulate = false;
    /// 保留计算后的直方图
    Mat b_hist, g_hist, r_hist;
    /// 计算B、G与R颜色通道的直方图分布:
    calcHist(&bgr_planes[0], 1, 0, Mat(), b_hist,
        1, &histSize, &histRange, uniform, accumulate);
    calcHist(&bgr_planes[1], 1, 0, Mat(), g_hist,
        1, &histSize, &histRange, uniform, accumulate);
    calcHist(&bgr_planes[2], 1, 0, Mat(), r_hist,
        1, &histSize, &histRange, uniform, accumulate);
    int hist_w = 512; int hist_h = 400;
    int bin_w = cvRound((double)hist_w / histSize);
    // 建立要显示直方图分布的图
    Mat histImage(hist_h, hist_w, CV_8UC3, Scalar(0, 0, 0));
    /// 将各颜色通道直方图分布结果归一化到 [ 0, histImage.rows ]范围
    normalize(b_hist, b_hist, 0, histImage.rows,
        NORM_MINMAX, -1, Mat());
    normalize(g_hist, g_hist, 0, histImage.rows,
```

```cpp
        NORM_MINMAX, -1, Mat());
    normalize(r_hist, r_hist, 0, histImage.rows,
        NORM_MINMAX, -1, Mat());
    /// 绘制每个颜色通道
    for (int i = 1; i < histSize; i++)
    {
        line(histImage, Point(bin_w*(i - 1),
            hist_h - cvRound(b_hist.at<float>(i - 1))),
            Point(bin_w*(i), hist_h - cvRound(b_hist.at<float>(i))),
            Scalar(255, 0, 0), 2, 8, 0);
        line(histImage, Point(bin_w*(i - 1),
            hist_h - cvRound(g_hist.at<float>(i - 1))),
            Point(bin_w*(i), hist_h - cvRound(g_hist.at<float>(i))),
            Scalar(0, 255, 0), 2, 8, 0);
        line(histImage, Point(bin_w*(i - 1),
            hist_h - cvRound(r_hist.at<float>(i - 1))),
            Point(bin_w*(i), hist_h - cvRound(r_hist.at<float>(i))),
            Scalar(0, 0, 255), 2, 8, 0);
    }
    /// 显示结果
    namedWindow("calcHist Demo", WINDOW_AUTOSIZE);
    imshow("calcHist Demo", histImage);
    waitKey(0);
    return 0;
}
```

程序说明

calcHist(const Mat* images, int nimages, const int* channels, InputArray mask, OutputArray hist, int dims, const int* histSize, const float** ranges, bool uniform=true, bool accumulate=false)：计算一组阵列的直方图分布

或是：

calcHist(const Mat* images, int nimages, const int* channels, InputArray mask, SparseMat& hist, int dims, const int* histSize, const float** ranges, bool uniform=true, bool accumulate=false)

（1）images：输入图像阵列。

（2）nimages：输入图像数量。

（3）channels：dims 的颜色通道清单，用来计算直方图分布。

（4）mask：可有可无的掩码。

（5）hist：输出的直方图分布。

（6）dims：直方图维度（Histogram dimensionality），必须是正值并且不可大于 CV_MAX_DIMS。

（7）histSize：每个维度的直方图大小阵列。

（8）ranges：每个维度直方图分布 bin 范围（bin boundaries）的直方图维度阵列的阵列。

（9）uniform：标示直方图分布是否一致的标志。

（10）accumulate：累积的标志（accumulation flag），如果设置，那么原直方图分布内容在函数内不会被清除。

执行结果如图 4-48 所示。

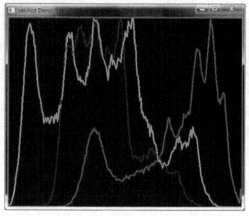

图 4-48

4.17　直方图分布比较

直方图分布的代码如下：

```
#include "opencv2/highgui/highgui.hpp"
#include "opencv2/imgproc/imgproc.hpp"
#include <iostream>
#include <stdio.h>
using namespace std;
using namespace cv;
int main(int argc, char** argv)
{
   Mat src_base, hsv_base;
   Mat src_test1, hsv_test1;
   Mat src_test2, hsv_test2;
   Mat hsv_half_down;
   src_base = imread("C:\\images\\hand_sample1.jpg", 1);
   src_test1 = imread("C:\\images\\hand_sample2.jpg", 1);
   src_test2 = imread("C:\\images\\hand_sample3.jpg", 1);
   /// 变换成 HSV 格式
   cvtColor(src_base, hsv_base, COLOR_BGR2HSV);
   cvtColor(src_test1, hsv_test1, COLOR_BGR2HSV);
   cvtColor(src_test2, hsv_test2, COLOR_BGR2HSV);
   /// 建立 HSV 格式基础图像一半的图像
   hsv_half_down = hsv_base(Range(hsv_base.rows / 2, hsv_base.rows - 1),
      Range(0, hsv_base.cols - 1));
   /// 用 30 bins 于色调(hue) and 32 于彩度(saturation)
   int h_bins = 50; int s_bins = 60;
   int histSize[] = { h_bins, s_bins };
   // 色调的范围 0 to 256, 彩度的范围 0 to 180
```

```cpp
    float s_ranges[] = { 0, 256 };
    float h_ranges[] = { 0, 180 };
    const float* ranges[] = { h_ranges, s_ranges };
    // 用第 0 与第 1 颜色通道(channels)
    int channels[] = { 0, 1 };
    /// 直方图分布图(Histograms)
    MatND hist_base;
    MatND hist_half_down;
    MatND hist_test1;
    MatND hist_test2;
    /// 为 HSV 图计算直方图分布
    calcHist(&hsv_base, 1, channels, Mat(),
        hist_base, 2, histSize, ranges, true, false);
    normalize(hist_base, hist_base, 0, 1,
        NORM_MINMAX, -1, Mat());
    calcHist(&hsv_half_down, 1, channels, Mat(),
        hist_half_down, 2, histSize, ranges, true, false);
    normalize(hist_half_down, hist_half_down,
        0, 1, NORM_MINMAX, -1, Mat());
    calcHist(&hsv_test1, 1, channels, Mat(),
        hist_test1, 2, histSize, ranges, true, false);
    normalize(hist_test1, hist_test1, 0, 1, NORM_MINMAX, -1, Mat());
    calcHist(&hsv_test2, 1, channels, Mat(),
        hist_test2, 2, histSize, ranges, true, false);
    normalize(hist_test2, hist_test2, 0, 1,
        NORM_MINMAX, -1, Mat());
    /// 采用直方图分布图比较法
    for (int i = 0; i < 4; i++)
    {
        int compare_method = i;
        double base_base = compareHist(hist_base,
            hist_base, compare_method);
        double base_half = compareHist(hist_base,
            hist_half_down, compare_method);
        double base_test1 = compareHist(hist_base,
            hist_test1, compare_method);
        double base_test2 = compareHist(hist_base,
            hist_test2, compare_method);
        printf(" Method [%d] Perfect, Base-Half, Base-Test(1), Base-Test(2)
            : %f, %f, %f, %f \n",
            i, base_base, base_half, base_test1, base_test2);
    }
    printf("Done \n");
    getchar();
    return 0;
}
```

程序说明

double compareHist(InputArray H1, InputArray H2, int method)　　直方图分布比较

或是

double compareHist(const SparseMat& H1, const SparseMat& H2, int method)

（1）H1：要比较的第一个直方图分布。

（2）H2：要比较的第二个直方图分布。

（3）method：比较方法。

- CV_COMP_CORREL：关联法（correlation）。
- CV_COMP_CHISQR：卡方法（chi-square）。
- CV_COMP_INTERSECT：交集法（intersection）。
- CV_COMP_BHATTACHARYYA Bhattacharyya 距离法（distance）。
- CV_COMP_HELLINGER：CV_COMP_BHATTACHARYYA 的同义字。

执行结果如图 4-49 和图 4-50 所示。

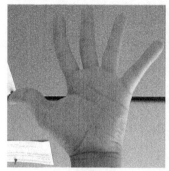 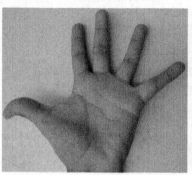 

（a）基础图像　　　　　（b）比较的图像 1　　　　　（c）比较的图像 2

图 4-49

图 4-50

4.18　反向投影

反向投影（Back Projection）是了解像素在直方图分布模式中分布的情况，可以计算特征的直方图分布来寻找图像的特征。

反向投影的代码如下:

```cpp
#include "opencv2/imgproc/imgproc.hpp"
#include "opencv2/highgui/highgui.hpp"
#include <iostream>
using namespace cv;
using namespace std;
/// 声明全局变量
Mat src; Mat hsv; Mat hue;
int bins = 25;
/// 声明函数
void Hist_and_Backproj(int, void*);
int main(int, char** argv)
{
    /// 加载图形文件
    src = imread("C:\\images\\hand_sample1.jpg", 1);
    /// 变换成 HSV 格式
    cvtColor(src, hsv, COLOR_BGR2HSV);
    /// 仅用色调(Hue)进行直方图分布
    hue.create(hsv.size(), hsv.depth());
    int ch[] = { 0, 0 };
    mixChannels(&hsv, 1, &hue, 1, ch, 1);
    const char* window_image = "Source image";
    namedWindow(window_image, WINDOW_AUTOSIZE);
    /// 建立滑杆以输入 bins
    createTrackbar("* Hue  bins: ", window_image,
        &bins, 180, Hist_and_Backproj);
    Hist_and_Backproj(0, 0);
    /// 显示结果
    imshow(window_image, src);
    /// 等待按键
    waitKey(0);
    return 0;
}
void Hist_and_Backproj(int, void*)
{
    MatND hist;
    int histSize = MAX(bins, 2);
    float hue_range[] = { 0, 180 };
    const float* ranges = { hue_range };
    /// 计算直方图分布(Histogram)
    calcHist(&hue, 1, 0, Mat(), hist, 1,
        &histSize, &ranges, true, false);
    /// 归一化(normalize)
    normalize(hist, hist, 0, 255,
        NORM_MINMAX, -1, Mat());
    /// 取得反向投影(Backprojection)
    MatND backproj;
    calcBackProject(&hue, 1, 0, hist,
        backproj, &ranges, 1, true);
    /// 显示反向投影
    imshow("BackProj", backproj);
    /// 绘制直方图分布图
    int w = 400; int h = 400;
    int bin_w = cvRound((double)w / histSize);
```

```
    Mat histImg = Mat::zeros(w, h, CV_8UC3);
    for (int i = 0; i < bins; i++)
    {
       rectangle(histImg, Point(i*bin_w, h),
           Point((i + 1)*bin_w,
           h - cvRound(hist.at<float>(i)*h / 255.0)),
           Scalar(0, 0, 255), -1);
    }
    /// 显示直方图分布结果
    imshow("Histogram", histImg);
}
```

程序说明

1. mixChannels(const Mat* src, size_t nsrcs, Mat* dst, size_t ndsts, const int* fromTo, size_t npairs)：复制输入图像的指定颜色通道到输出图像

　　或是：

mixChannels(const vector<Mat>& src, vector<Mat>& dst, const int* fromTo, size_t npairs)

（1）src：输入图像或矩阵向量，必须是同大小与景深。

（2）nsrcs：src 的矩阵数。

（3）dst：输出图像或矩阵向量。

（4）ndsts：dst 的矩阵数。

（5）fromTo：索引对（index pairs）的阵列，说明颜色通道是从何处复制到何处。

（6）npairs：fromTo 索引对数量。

2. calcBackProject(const Mat* images, int nimages, const int* channels, InputArray hist, OutputArray backProject, const float** ranges, double scale=1, bool uniform=true)：计算直方图分布的反向投影

　　或是：

calcBackProject(const Mat* images, int nimages, const int* channels, const SparseMat& hist, OutputArray backProject, const float** ranges, double scale=1, bool uniform=true)

（1）images：输入图像或矩阵向量，必须是同大小与景深，但各自有不同的颜色通道。

（2）nimages：输入图像的数量。

（3）channels：计算反向投影的颜色通道清单，其数量必须与直方图分布维度相同。

（4）hist：输入的直方图分布。

（5）backProject：反向投影结果单颜色通道阵列，都是同大小与景深。

（6）ranges：每个维度直方图分布 bin 范围（bin boundaries）的直方图维度阵列的阵列。

（7）scale：反向投影结果的缩放因子，为可选项参数。

（8）uniform：标示直方图分布是否一致的标志。

执行结果如图 4-51 所示。

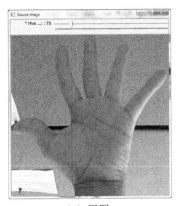
（a）原图

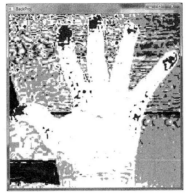
（b）反向投影

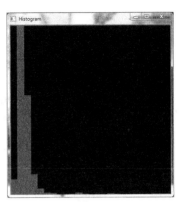
（c）直方图分布

图 4-51

酷炫范例

如下是一个示例，执行结果如图 4-52 所示。

```cpp
#include "opencv2/imgproc/imgproc.hpp"
#include "opencv2/highgui/highgui.hpp"
#include <iostream>
using namespace cv;
using namespace std;
/// 声明全局变量
Mat src; Mat hsv;
Mat mask;
int lo = 20; int up = 20;
const char* window_image = "Source image";
/// 声明函数
void Hist_and_Backproj();
void pickPoint(int event, int x, int y, int, void*);
int main(int, char** argv)
{
    /// 加载图形文件
    src = imread("C:\\images\\hand_sample1.jpg", 1);
    /// 变换成 HSV 格式
    cvtColor(src, hsv, COLOR_BGR2HSV);
    /// 显示图形文件
    namedWindow(window_image, WINDOW_AUTOSIZE);
    imshow(window_image, src);
    /// 建立滑杆进行填满式阈值处理(floodfill thresholds)
    createTrackbar("Low thresh", window_image, &lo, 255, 0);
    createTrackbar("High thresh", window_image, &up, 255, 0);
    /// 设置鼠标回调函数
    setMouseCallback(window_image, pickPoint, 0);
    waitKey(0);
    return 0;
}
```

```cpp
void pickPoint(int event, int x, int y, int, void*)
{
    /// 没有按下鼠标左键则跳出
    if (event != EVENT_LBUTTONDOWN)
        return;
    Point seed = Point(x, y);
    // 取得掩码并填满
    int newMaskVal = 255;
    Scalar newVal = Scalar(120, 120, 120);
    int connectivity = 8;
    int flags = connectivity + (newMaskVal << 8) +
        FLOODFILL_FIXED_RANGE + FLOODFILL_MASK_ONLY;
    Mat mask2 = Mat::zeros(src.rows + 2, src.cols + 2, CV_8UC1);
    floodFill(src, mask2, seed, newVal, 0,
        Scalar(lo, lo, lo), Scalar(up, up, up), flags);
    mask = mask2(Range(1, mask2.rows - 1),
        Range(1, mask2.cols - 1));
    imshow("Mask", mask);
    Hist_and_Backproj();
}
void Hist_and_Backproj()
{
    MatND hist;
    int h_bins = 30; int s_bins = 32;
    int histSize[] = { h_bins, s_bins };
    float h_range[] = { 0, 179 };
    float s_range[] = { 0, 255 };
    const float* ranges[] = { h_range, s_range };
    int channels[] = { 0, 1 };
    /// 取得直方图分布
    calcHist(&hsv, 1, channels, mask, hist, 2,
        histSize, ranges, true, false);
    /// 归一化
    normalize(hist, hist, 0, 255, NORM_MINMAX, -1, Mat());
    /// 反向投影
    MatND backproj;
    calcBackProject(&hsv, 1, channels, hist,
        backproj, ranges, 1, true);
    /// 显示反向投影结果
    imshow("BackProj", backproj);
}
```

程序说明

int floodFill(InputOutputArray image, Point seedPoint, Scalar newVal, Rect* rect=0, Scalar loDiff=Scalar(), Scalar upDiff=Scalar(), int flags=4)：用颜色填满连接的组件(component)

或是：

int floodFill(InputOutputArray image, InputOutputArray mask, Point seedPoint, Scalar newVal, Rect* rect=0, Scalar loDiff=Scalar(), Scalar upDiff=Scalar(), int flags=4)

（1）image：输入与输出图像。

（2）seedPoint：开始点。
（3）newVal：重绘区域像素的新值。
（4）rect：输出参数为可选项，重绘区域的最小矩形区。
（5）loDiff：组件附近已观察像素间最大较低亮度或色差，或是加到组件的种子像素（seed pixel）。
（6）upDiff：组件附近已观察像素间最大较高亮度或色差，或是加到组件的种子像素（seed pixel）。
（7）flags：

- FLOODFILL_FIXED_RANGE 如果设置就是目前像素与种子像素的色差，则就是目前像素与附近像素间色差。
- FLOODFILL_MASK_ONLY 如果设置此函数不变更输入图像，newVal 就被忽略。

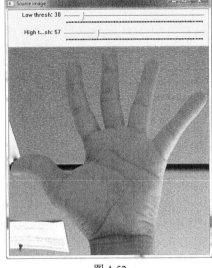

图 4-52

（8）mask：掩码。

先使用滑杆设置阈值，再用鼠标在食指上点一下，结果如图 4-53 所示。鼠标的说明请参考 11.2 节。

 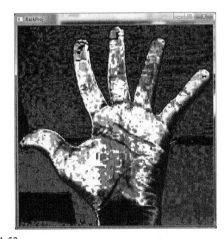

图 4-53

图像着色

给图像着色的代码如下：

```
#include <opencv2/core/core.hpp>
#include <opencv2/highgui/highgui.hpp>
#include <opencv2/video/video.hpp>
#include <opencv2/imgproc/imgproc.hpp>
#include <iostream>
```

```cpp
using namespace cv;
using namespace std;
// 声明全局变量
Mat image0, image, gray, mask;
int ffillMode = 1;      // 填补着色模式
int loDiff = 20, upDiff = 20; // 滑杆起始，调整填色域
int connectivity = 4;   // connectivity 模式
int isColor = true;     // 灰度或彩色
bool useMask = false;   // 使用遮蔽
int newMaskVal = 255;
// 鼠标回调函数
static void onMouse(int event, int x, int y, int, void*)
{
    if (event != EVENT_LBUTTONDOWN)
        return;
    Point seed = Point(x, y);
    int lo = ffillMode == 0 ? 0 : loDiff;
    int up = ffillMode == 0 ? 0 : upDiff;
    int flags = connectivity + (newMaskVal << 8) +
        (ffillMode == 1 ? FLOODFILL_FIXED_RANGE : 0);
    int b = (unsigned)theRNG() & 255;
    int g = (unsigned)theRNG() & 255;
    int r = (unsigned)theRNG() & 255;
    Rect ccomp;
    Scalar newVal = isColor ? Scalar(b, g, r) : Scalar(r*0.299 + g*0.587 +
                b*0.114);
    Mat dst = isColor ? image : gray;
    int area;
    // 使用遮蔽
    if (useMask)
    {
        threshold(mask, mask, 1, 128, THRESH_BINARY);
        area = floodFill(dst, mask, seed, newVal, &ccomp, Scalar(lo, lo, lo),
           Scalar(up, up, up), flags);
        imshow("mask", mask);
    }
    else
    {
        area = floodFill(dst, seed, newVal, &ccomp, Scalar(lo, lo, lo),
           Scalar(up, up, up), flags);
    }
    imshow("image", dst);
    cout << area << " 像素(pixel)被重绘\n";
}
int main(int argc, char** argv)
{
    char* filename = "C:\\images\\fruits.jpg";
    image0 = imread(filename, 1);
    if (image0.empty())
    {
     cout << "读取文档失败\n";
     return 0;
    }
    image0.copyTo(image);
    cvtColor(image0, gray, COLOR_BGR2GRAY);
    mask.create(image0.rows + 2, image0.cols + 2, CV_8UC1);
    namedWindow("image", 0);
```

```cpp
// 建立窗口滑杆回调功能
createTrackbar("lo_diff", "image", &loDiff, 255, 0);
createTrackbar("up_diff", "image", &upDiff, 255, 0);
// 建立鼠标回调功能
setMouseCallback("image", onMouse, 0);
for (;;)
{
    imshow("image", isColor ? image : gray);
    int c = waitKey(0);
    if ((c & 255) == 27)
    {
        cout << "结束 ...\n";
        break;
    }
    switch ((char)c)
    {
    case 'c':
        if (isColor)
        {
            cout << "设成灰度\n";
            cvtColor(image0, gray, COLOR_BGR2GRAY);
            mask = Scalar::all(0);
            isColor = false;
        }
        else
        {
            cout << "设成彩色\n";
            image0.copyTo(image);
            mask = Scalar::all(0);
            isColor = true;
        }
        break;
    case 'm':
        if (useMask)
        {
            destroyWindow("mask");
            useMask = false;
        }
        else
        {
            namedWindow("mask", 0);
            mask = Scalar::all(0);
            imshow("mask", mask);
            useMask = true;
        }
        break;
    case 'r':
        cout << "还原图像\n";
        image0.copyTo(image);
        cvtColor(image, gray, COLOR_BGR2GRAY);
        mask = Scalar::all(0);
        break;
    case 's':
        cout << "设置简易填补着色模式\n";
        ffillMode = 0;
        break;
```

```
            case 'f':
                cout << "设置固定范围填补着色模式\n";
                ffillMode = 1;
                break;
            case 'g':
                cout << "设置 Gradient (floating range) 填补着色模式\n";
                ffillMode = 2;
                break;
            case '4':
                cout << "设置 4-connectivity 模式\n";
                connectivity = 4;
                break;
            case '8':
                cout << "设置 8-connectivity 模式\n";
                connectivity = 8;
                break;
        }
    }
    return 0;
}
```

本程序是 floodFill 函数的实际应用。

执行结果如图 4-54 和图 4-55 所示。

图 4-54

图 4-55

经典范例

```cpp
#include <opencv2/video/tracking.hpp>
#include <opencv2/imgproc/imgproc.hpp>
#include <opencv2/video/video.hpp>
#include <opencv2/core/core.hpp>
#include <opencv2/highgui/highgui.hpp>
#include <iostream>
#include <ctype.h>
using namespace cv;
using namespace std;
Mat image;
// 按键操作的起始值
bool backprojMode = false;
bool selectObject = false;
int trackObject = 0;
bool showHist = true;
// 鼠标选取范围
Point origin;
Rect selection;
// 颜色范围起始值
int vmin = 10, vmax = 256, smin = 30;
static void onMouse(int event, int x, int y, int, void*)
{
    if (selectObject)
    {
        selection.x = MIN(x, origin.x);
        selection.y = MIN(y, origin.y);
        selection.width = std::abs(x - origin.x);
        selection.height = std::abs(y - origin.y);
        selection &= Rect(0, 0, image.cols, image.rows);
    }
    switch (event)
    {
    case EVENT_LBUTTONDOWN:
        origin = Point(x, y);
        selection = Rect(x, y, 0, 0);
        selectObject = true;
        break;
     case EVENT_LBUTTONUP:
        selectObject = false;
        if (selection.width > 0 && selection.height > 0)
            trackObject = -1;
        break;
    }
}
static void help()
{
    cout << "\n\n热键: \n"
        "\tESC - 结束程序\n"
        "\t c  - 停止追踪\n"
        "\t b  - 反向投影切换\n"
        "\t h  - 显示或隐藏所选的直方图分布\n"
        "\t p  - 拍摄暂停切换\n"
        "\n 用鼠标选取范围开始追踪\n";
```

```cpp
}
const char* keys =
{
    // @代号 | 值 | 说明
    "{@camera_nbr| 0 | camera number}"
};
int main(int argc, const char** argv)
{
    help();
    VideoCapture cap;
    Rect trackWindow;
    int hsize = 16;
    float hranges[] = { 0, 180 };
    const float* phranges = hranges;
    // 取得相机代号
    CommandLineParser parser(argc, argv, keys);
    int camNum = parser.get<int>("camera_nbr");
    cap.open(camNum);
    if (!cap.isOpened())
    {
        help();
        cout << "*** 无法启动相机 ***\n";
        cout << "目前参数值: \n";
        parser.printParams();
        getchar();
        return -1;
    }
    namedWindow("Histogram", CV_WINDOW_AUTOSIZE);
    namedWindow("CamShift", CV_WINDOW_NORMAL);
    setMouseCallback("CamShift", onMouse, 0);
    // 用滑杆指定颜色范围值
    createTrackbar("Vmin", "CamShift", &vmin, 256, 0);
    createTrackbar("Vmax", "CamShift", &vmax, 256, 0);
    createTrackbar("Smin", "CamShift", &smin, 256, 0);
    Mat frame, hsv, hue, mask, hist,
        histimg = Mat::zeros(200, 320, CV_8UC3), backproj;
    bool paused = false;
    for (;;)
    {
        if (!paused)
        {
            cap >> frame;
            if (frame.empty())
                break;
        }
        frame.copyTo(image);
        if (!paused)
        {
            cvtColor(image, hsv, COLOR_BGR2HSV);
            if (trackObject)
            {
                // 只将hsv于两色间的值存入mask内
                inRange(hsv, Scalar(0, smin, MIN(vmin, vmax)),
                    Scalar(180, 256, MAX(vmin, vmax)), mask);
                int ch[] = { 0, 0 };
```

```cpp
            hue.create(hsv.size(), hsv.depth());
            // 从 hsv 复制指定的颜色通道数到 hue
            mixChannels(&hsv, 1, &hue, 1, ch, 1);
            if (trackObject < 0)
            {
                Mat roi(hue, selection),
                    maskroi(mask, selection);
                // 直方图分布
                calcHist(&roi, 1, 0, maskroi,
                    hist, 1, &hsize, &phranges);
                // 归一化
                normalize(hist, hist, 0, 255, NORM_MINMAX);
                trackWindow = selection;
                trackObject = 1;
                histimg = Scalar::all(0);
                int binW = histimg.cols / hsize;
                Mat buf(1, hsize, CV_8UC3);
                for (int i = 0; i < hsize; i++)
                    buf.at<Vec3b>(i) = Vec3b(saturate_cast<uchar>
                    (i*180. / hsize), 255, 255);
                cvtColor(buf, buf, COLOR_HSV2BGR);
                // 绘制直方图分布
                for (int i = 0; i < hsize; i++)
                {
                  int val = saturate_cast<int>
                      (hist.at<float>(i) * histimg.rows / 255);
                  rectangle(histimg,
                    Point(i*binW, histimg.rows),
                    Point((i + 1)*binW, histimg.rows - val),
                    Scalar(buf.at<Vec3b>(i)), -1, 8);
                }
            }
            // 反向投影
            calcBackProject(&hue, 1, 0, hist, backproj, &phranges);
            backproj &= mask;
            // 寻找反向投影的中心、大小与方向(orientation)
            RotatedRect trackBox = CamShift(backproj, trackWindow,
                TermCriteria(TermCriteria::EPS | TermCriteria::COUNT, 10, 1));
            if (trackWindow.area() <= 1)
            {
              int cols = backproj.cols,
                 rows = backproj.rows,
                 r = (MIN(cols, rows) + 5) / 6;
              trackWindow = Rect(trackWindow.x - r,
                 trackWindow.y - r, trackWindow.x + r,
                 trackWindow.y + r) & Rect(0, 0, cols, rows);
            }
            // 将反向投影放入窗口图像
            if (backprojMode)
                cvtColor(backproj, image, COLOR_GRAY2BGR);
            // 绘制出追踪范围
            ellipse(image, trackBox, Scalar(0, 0, 255), 3, CV_AA);
        }
    }
    else if (trackObject < 0)
```

```
            paused = false;
        // 将鼠标选取范围的每个位反转
        if (selectObject && selection.width > 0
            && selection.height > 0)
        {
            Mat roi(image, selection);
            bitwise_not(roi, roi);
        }
        imshow("CamShift", image);
        if (showHist)
            imshow("Histogram", histimg);
        char c = (char)waitKey(10);
        if (c == 27)
            break;
        switch (c)
        {
        case 'b':
            backprojMode = !backprojMode;
            break;
        case 'c':
            trackObject = 0;
            histimg = Scalar::all(0);
            break;
        case 'h':
            showHist = !showHist;
            if (!showHist)
                destroyWindow("Histogram");
            else
                namedWindow("Histogram", 1);
            break;
        case 'p':
            paused = !paused;
            break;
        default:
            ;
        }
    }
    return 0;
}
```

执行结果如图 4-56 所示。

图 4-56

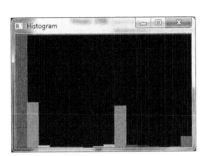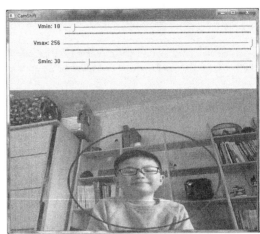

图 4-56（续）

4.19 模板匹配

模板匹配（Template Matching）是在图中找到与模板吻合的地方。其比对方式是将模板从左上角开始，由左至右且由上到下逐个像素开始移动匹配，将匹配结果分为好、坏或相似。

具体代码如下：

```
#include "opencv2/highgui/highgui.hpp"
#include "opencv2/imgproc/imgproc.hpp"
#include <iostream>
#include <stdio.h>
using namespace std;
using namespace cv;
/// 声明全局变量
Mat img; Mat templ; Mat result;
const char* image_window = "Source Image";
const char* result_window = "Result window";
int match_method;
// 滑杆最大值
int max_Trackbar = 5;
/// 声明函数
void MatchingMethod(int, void*);
int main(int, char** argv)
{
    /// 加载图形文件
    img = imread("C:\\images\\lena.jpg", 1);
    // 加载模板
    templ = imread("C:\\images\\crop-lena.jpg", 1);
    /// 建立窗口
    namedWindow(image_window, WINDOW_AUTOSIZE);
    namedWindow(result_window, WINDOW_AUTOSIZE);
    /// 建立滑杆
```

```cpp
        // 0: SQDIFF              1: SQDIFF NORMED   2: TM CCORR
        // 3: TM CCORR NORMED  4: TM COEFF          5: TM COEFF NORMED
        const char* trackbar_label = "Method: ";
        createTrackbar(trackbar_label, image_window,
            &match_method, max_Trackbar, MatchingMethod);
        // 匹配
        MatchingMethod(0, 0);
        waitKey(0);
        return 0;
}
// 因为要传递参数但又不会使用到
void MatchingMethod(int, void*)
{
        /// 复制原图
        Mat img_display;
        img.copyTo(img_display);
        /// 建立模板于原图匹配结果的矩阵
        int result_cols = img.cols - templ.cols + 1;
        int result_rows = img.rows - templ.rows + 1;
        result.create(result_cols, result_rows, CV_32FC1);
        /// 执行匹配与归一化
        matchTemplate(img, templ, result, match_method);
        normalize(result, result, 0, 1, NORM_MINMAX, -1, Mat());
        double minVal, maxVal;
        Point minLoc, maxLoc, matchLoc;
        ///寻找图中最大与最小值及其位置
        minMaxLoc(result, &minVal, &maxVal, &minLoc, &maxLoc, Mat());
        /// SQDIFF 与 SQDIFF_NORMED 是值越低，比较结果越好
        /// 其他方法是越高越好
        if (match_method == CV_TM_SQDIFF
            || match_method == CV_TM_SQDIFF_NORMED)
        {
         matchLoc = minLoc;
        }
        else
        {
         matchLoc = maxLoc;
        }
        // 绘制匹配结果范围
        // 原图
        rectangle(img_display, matchLoc,
            Point(matchLoc.x + templ.cols, matchLoc.y + templ.rows),
            Scalar::all(0), 2, 8, 0);
        // 归一化结果
        rectangle(result, matchLoc,
            Point(matchLoc.x + templ.cols, matchLoc.y + templ.rows),
            Scalar::all(0), 2, 8, 0);
        /// 显示结果
        imshow(image_window, img_display);
        imshow(result_window, result);
        return;
}
```

程序说明

1. matchTemplate(InputArray image, InputArray templ, OutputArray result, int method)：比较模板与图像重叠区域

    （1）image：要比较的图像。
    （2）templ：要寻找的模板。
    （3）result：比较结果。
    （4）method：指定比较的方式。
    CV_TM_SQDIFF

    $$R(x,y) = \sum_{x',y'}(T(x',y') - I(x+x', y+y'))^2$$

    CV_TM_SQDIFF_NORMED

    $$R(x,y) = \frac{\sum_{x',y'}(T(x',y') - I(x+x', y+y'))^2}{\sum_{x',y'}T(x',y')^2 \cdot \sum_{x',y'}I(x+x', y+y')^2}$$

    CV_TM_CCORR

    $$R(x,y) = \sum_{x',y'}(T(x',y') \cdot I(x+x', y+y'))$$

    CV_TM_CCORR_NORMED

    $$R(x,y) = \frac{\sum_{x',y'}(T(x',y') \cdot I(x+x', y+y'))^2}{\sqrt{\sum_{x',y'}T(x',y')^2 \cdot \sum_{x',y'}I(x+x', y+y')^2}}$$

    CV_TM_CCOEFF

    $$R(x,y) = \sum_{x',y'}(T'(x',y') - I'(x+x', y+y'))$$

    $$T'(x',y') = T(x',y') - 1/(w \cdot h) \cdot \sum_{x'',y''}T(x'',y'')$$

    $$I'(x+x',y+y') = I(x+x',y+y') - 1/(w \cdot h) \cdot \sum_{x'',y''}I(x+x'',y+y'')$$

    CV_TM_CCOEFF_NORMED

    $$R(x,y) = \frac{\sum_{x',y'}(T(x',y') \cdot I(x+x', y+y'))}{\sum_{x',y'}T'(x',y')^2 \cdot \sum_{x',y'}I'(x+x', y+y')^2}$$

2. minMaxLoc(InputArray src, double* minVal, double* maxVal=0, Point* minLoc=0, Point* maxLoc=0, InputArray mask=noArray())：寻找图中最大值与最小值

或是：

minMaxLoc(const SparseMat& a, double* minVal, double* maxVal, int* minIdx=0, int* maxIdx=0)

（1）src：输入单颜色通道图像。
（2）minVal：返回的最小值，若设为 NULL 表示不需要。
（3）maxVal：返回的最大值，若设为 NULL 表示不需要。
（4）minLoc：返回图像中最小位置，若设为 NULL 表示不需要。
（5）maxLoc：返回图像中最大位置，若设为 NULL 表示不需要。
（6）mask：掩码，参数可选，用来选择次图（sub-array）。

执行结果如图 4-57 所示。

（a）原图　　　　　　　　　　　　（b）模板图

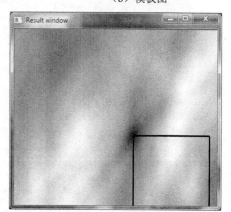

（c）匹配结果　　　　　　　　　　（d）归一化结果

图 4-57

如果你阅读完 11.1 节并了解图像的秘密，任何原图或模板是有藏密的图，那模板比对还能正确匹配结果吗？

4.20 寻找图的轮廓

寻找图的轮廓的代码如下：

```cpp
#include "opencv2/highgui/highgui.hpp"
#include "opencv2/imgproc/imgproc.hpp"
#include <iostream>
#include <stdio.h>
#include <stdlib.h>
using namespace cv;
using namespace std;
/// 声明全局变量
Mat src; Mat src_gray;
int thresh = 100;
int max_thresh = 255;
RNG rng(12345);
/// 声明函数
void thresh_callback(int, void*);
int main(int, char** argv)
{
    /// 加载图形文件
    src = imread("C:\\images\\HappyFish.jpg", 1);
    /// 转成灰度
    cvtColor(src, src_gray, COLOR_BGR2GRAY);
    /// 模糊处理
    blur(src_gray, src_gray, Size(3, 3));
    /// 建立窗口
    const char* source_window = "Source";
    namedWindow(source_window, WINDOW_AUTOSIZE);
    imshow(source_window, src);
    /// 建立滑杆
    createTrackbar(" Canny thresh:", "Source",
        &thresh, max_thresh, thresh_callback);
    thresh_callback(0, 0);
    waitKey(0);
    return(0);
}
void thresh_callback(int, void*)
{
    Mat canny_output;
    vector<vector<Point> > contours;
    vector<Vec4i> hierarchy;
    /// 使用 canny 检测边缘
    Canny(src_gray, canny_output, thresh, thresh * 2, 3);
    imshow("canny", canny_output);
    /// 寻找轮廓(contours)
    findContours(canny_output, contours, hierarchy,
        RETR_TREE, CHAIN_APPROX_SIMPLE, Point(0, 0));
    imshow("findContour", canny_output);
    Mat drawing = Mat::zeros(canny_output.size(), CV_8UC3);
```

```
        for (size_t i = 0; i< contours.size(); i++)
        {
            Scalar color = Scalar(rng.uniform(0, 255),
                rng.uniform(0, 255), rng.uniform(0, 255));
            /// 绘制轮廓
            drawContours(drawing, contours, (int)i,
                color, 2, 8, hierarchy, 0, Point());
        }
        /// 显示结果
        namedWindow("Contours", WINDOW_AUTOSIZE);
        imshow("Contours", drawing);
    }
```

程序说明

1. findContours(InputOutputArray image, OutputArrayOfArrays contours, OutputArray hierarchy, int mode, int method, Point offset=Point())：寻找二元图像的轮廓

   或是：

   findContours(InputOutputArray image, OutputArrayOfArrays contours, int mode, int method, Point offset=Point())

   （1）image：输入图像。
   （2）contours：检测的轮廓，轮廓是以点的向量存储。
   （3）hierarchy：可有可无的输出向量，内容为输入图的拓扑（topology）信息。
   （4）mode：取得轮廓的模式。
   - CV_RETR_EXTERNAL：只取最外层轮廓，并设置hieraechy[i][2]=hierarchy[i][3]=-1。
   - CV_RETR_LIST：取得所有轮廓，并不建立层级（hierarchy）关系。
   - CV_RETR_CCOMP：取得所有轮廓，并组织成两层级结构（two-level hierarchy）。
   - CV_RETR_TREE：取得所有轮廓，并重建嵌套式轮廓（nested contours）的全层级（full hierarchy）。

   （5）method：取得轮廓的方法。
   - CV_CHAIN_APPROX_NONE：存储所有轮廓点。
   - CV_CHAIN_APPROX_SIMPLE：压缩水平、垂直、对角而只留下末点。

   （6）offset：是可有可无的每个轮廓点的位移量。

2. drawContours(InputOutputArray image, InputArrayOfArrays contours, int contourIdx, const Scalar& color, int thickness=1, int lineType=8, InputArray hierarchy=noArray(), int maxLevel=INT_MAX, Point offset=Point())：绘制轮廓

   （1）image：输出图像。
   （2）contours：所有输入的轮廓。
   （3）contourIdx：要绘制的轮廓。

（4）color：轮廓绘制的颜色。
（5）thickness：轮廓绘制的厚度。
（6）lineType：轮廓线条的形态。
（7）hierarchy：是可有可无的阶层信息。
（8）maxLevel：绘制最大阶层的轮廓。
- 如果值为 0，则绘制指定阶层的轮廓。
- 如果值为 1，则绘制所有嵌套的轮廓。
- 如果值为 2，则绘制所有嵌套的轮廓。

（9）offset：是可有可无的轮廓偏移参数。

执行结果如图 4-58 所示。

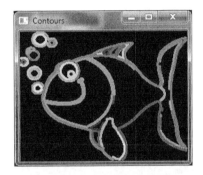

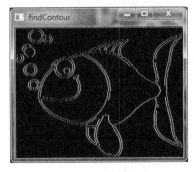

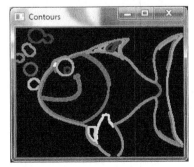

图 4-58

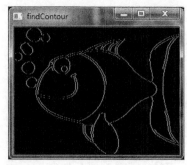

图 4-58（续）

以轮廓寻找车牌

以轮廓寻找车牌的代码如下：

```cpp
#include "opencv2/core/core.hpp"
#include "opencv2/imgproc/imgproc.hpp"
#include "opencv2/highgui/highgui.hpp"
#include <iostream>
#include <math.h>
#include <string.h>
using namespace cv;
using namespace std;
int thresh = 50, N = 11;
const char* wndname = "Square Detection Demo";
// 在两个向量间寻找余弦角
// pt0->pt1 与 pt0->pt2
static double angle( Point pt1, Point pt2, Point pt0 )
{
    double dx1 = pt1.x - pt0.x;
    double dy1 = pt1.y - pt0.y;
    double dx2 = pt2.x - pt0.x;
    double dy2 = pt2.y - pt0.y;
    return (dx1*dx2 + dy1*dy2)/sqrt((dx1*dx1 + dy1*dy1)
            *(dx2*dx2 + dy2*dy2) + 1e-10);
}
// 函数将检测到的方形存储在 squares 内
static void findSquares( const Mat& image,
        vector<vector<Point> >& squares )
{
    squares.clear();
    Mat pyr, timg, gray0(image.size(), CV_8U), gray;
    // 通过缩放来去除噪声
    pyrDown(image, pyr, Size(image.cols/2, image.rows/2));
    pyrUp(pyr, timg, image.size());
    vector<vector<Point> > contours;
    // 在每个颜色通道寻找四边形
    for( int c = 0; c < 3; c++ )
    {
        int ch[] = {c, 0};
        mixChannels(&timg, 1, &gray0, 1, ch, 1);
        // 测试每种临界等级(threshold levels)
        for( int l = 0; l < N; l++ )
```

```cpp
        {
            if( l == 0 )
            {
                Canny(gray0, gray, 0, thresh, 5);
                // 用膨胀去除边缘的空洞
                dilate(gray, gray, Mat(), Point(-1,-1));
            }
            else
            {
                // 使用阈值
                // tgray(x,y) = gray(x,y) < (l+1)*255/N ? 255 : 0
                gray = gray0 >= (l+1)*255/N;
            }
            // 将找到的轮廓存储在清单内
            findContours(gray, contours, CV_RETR_LIST,
                         CV_CHAIN_APPROX_SIMPLE);
            vector<Point> approx;
            // 测试每个轮廓
            for( size_t i = 0; i < contours.size(); i++ )
            {
                // 用轮廓周边(contour perimeter)正确等比(accuracy
                // proportional)取得近似轮廓(approximate contour)
                approxPolyDP(Mat(contours[i]), approx,
                             arcLength(Mat(contours[i]), true)*0.02, true);
                // 近似四方型轮廓应该具有四个顶点并且是凸面的(convex)
                // 不考虑轮廓区域的转向(orientation),所以取绝对值
                if( approx.size() == 4 &&
                    fabs(contourArea(Mat(approx))) > 1000 &&
                    isContourConvex(Mat(approx)) )
                {
                    double maxCosine = 0;
                    for( int j = 2; j < 5; j++ )
                    {
                        // 在连接的两个边寻找最大的余弦角
                        double cosine = fabs(angle(approx[j%4],
                                approx[j-2], approx[j-1]));
                        maxCosine = MAX(maxCosine, cosine);
                    }
                    // 如果所有余弦角很小就将方形四个顶点写入结果序列
                    // 搜索到的四边形应该是近似直角 90 度
                    if( maxCosine < 0.3 )
                        squares.push_back(approx);
                }
            }
        }
    }
}
// 绘制四边形
static void drawSquares( Mat& image,
        const vector<vector<Point> >& squares )
{
    for( size_t i = 0; i < squares.size(); i++ )
    {
        const Point* p = &squares[i][0];
```

```
            int n = (int)squares[i].size();
            polylines(image, &p, &n, 1, true,
                    Scalar(0,255,0), 3, CV_AA);
    }
    imshow(wndname, image);
}
int main(int argc, char** argv)
{
    char* names = "c:/images/2998274.jpg";
    namedWindow( wndname, 1 );
    vector<vector<Point> > squares;
    Mat image = imread(names, 1);
    if( image.empty() )
    {
        cout << "Couldn't load " << names << endl;
        return 1;
    }
    findSquares(image, squares);
    drawSquares(image, squares);
    waitKey(0);
    return 0;
}
```

程序说明

approxPolyDP(InputArray curve, OutputArray approxCurve, double epsilon, bool closed)：指定精确度接近的（approximates）多边曲线

（1）curve：输入的二维向量点。

（2）approxCurve：近似的多边曲线结果。

（3）epsilon：精确度。

（4）closed：若值为true，则曲线头尾相连，否则不相连。

因为车牌是长方形的，所以最适宜用轮廓来寻找，而且也没有车牌大小与车牌拍照距离的问题。首先要找到车牌，然后才是识别车牌，如图4-59所示。

图 4-59

图 4-59（续）

4.21 凸包

凸包的代码如下：

```
#include "opencv2/highgui/highgui.hpp"
#include "opencv2/imgproc/imgproc.hpp"
#include <iostream>
#include <stdio.h>
#include <stdlib.h>
using namespace cv;
using namespace std;
/// 声明全局变量
Mat src; Mat src_gray;
int thresh = 100;
int max_thresh = 255;
RNG rng(12345);
/// 声明函数
void thresh_callback(int, void*);
int main(int, char** argv)
{
    /// 加载图形文件
    src = imread("C:\\images\\hand_sample1.jpg", 1);
    /// 转成灰度
    cvtColor(src, src_gray, COLOR_BGR2GRAY);
    /// 模糊
    blur(src_gray, src_gray, Size(3, 3));
    /// 建立窗口
    const char* source_window = "Source";
    namedWindow(source_window, WINDOW_AUTOSIZE);
    imshow(source_window, src);
    /// 建立滑杆
    createTrackbar(" Threshold:", "Source",
        &thresh, max_thresh, thresh_callback);
    thresh_callback(0, 0);
    waitKey(0);
    return(0);
```

```cpp
}
void thresh_callback(int, void*)
{
    Mat src_copy = src.clone();
    Mat threshold_output;
    vector<vector<Point> > contours;
    vector<Vec4i> hierarchy;
    /// 使用阈值(Threshold)检测边缘
    threshold(src_gray, threshold_output,
     thresh, 255, THRESH_BINARY);
    /// 寻找轮廓(contour)
    findContours(threshold_output, contours, hierarchy,
       RETR_TREE, CHAIN_APPROX_SIMPLE, Point(0, 0));
    /// 为每个轮廓寻找凸包(convex hull) 对象
    vector<vector<Point> >hull(contours.size());
    for (size_t i = 0; i < contours.size(); i++)
    {
        convexHull(Mat(contours[i]), hull[i], false);
    }
    /// 绘轮廓与凸包
    Mat drawing = Mat::zeros(threshold_output.size(), CV_8UC3);
    for (size_t i = 0; i< contours.size(); i++)
    {
        Scalar color = Scalar(rng.uniform(0, 255),
          rng.uniform(0, 255), rng.uniform(0, 255));
        /// 绘制轮廓
        drawContours(drawing, contours, (int)i, color,
          1, 8, vector<Vec4i>(), 0, Point());
        /// 绘制凸包
        drawContours(drawing, hull, (int)i, color, 1,
          8, vector<Vec4i>(), 0, Point());
    }
    /// 显示结果
    namedWindow("Hull demo", WINDOW_AUTOSIZE);
    imshow("Hull demo", drawing);
}
```

程序说明

convexHull(InputArray points, OutputArray hull, bool clockwise=false, bool returnPoints=true)：寻找设置点的凸包。

（1）points：输入的二维设置点。

（2）hull：输出的凸包向量点。

（3）clockwise：转向标志，若值为 true，则顺时针转，否则就是反时针转。

（4）returnPoints：操作标志，若为 true，则回传凸包的向量点，否则返回凸包点的指标（indices）。

执行结果如图 4-60 所示。

第 4 章 ImgProc 模块

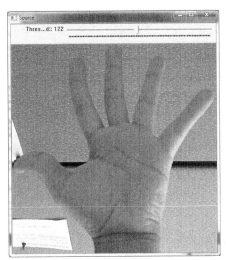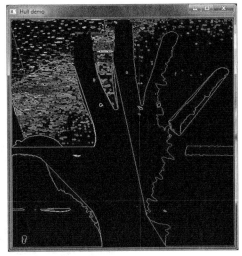

图 4-60

凸包示范

凸包示范的代码如下:

```
#include "opencv2/imgproc/imgproc.hpp"
#include "opencv2/highgui/highgui.hpp"
#include <fstream>
#include <iostream>
using namespace cv;
using namespace std;
int main( int /*argc*/, char** /*argv*/ )
{
    Mat img(500, 500, CV_8UC3);
    RNG& rng = theRNG();
    for(;;)
    {
        char key;
        int i, count = (unsigned)rng%100 + 1;
        vector<Point> points;
        for( i = 0; i < count; i++ )
        {
            Point pt;
            pt.x = rng.uniform(img.cols/4, img.cols*3/4);
            pt.y = rng.uniform(img.rows/4, img.rows*3/4);
            points.push_back(pt);
        }
        vector<int> hull;
        convexHull(Mat(points), hull, true);
        img = Scalar::all(0);
        for( i = 0; i < count; i++ )
            circle(img, points[i], 3,
         Scalar(0, 0, 255), CV_FILLED, CV_AA);
        int hullcount = (int)hull.size();
        Point pt0 = points[hull[hullcount-1]];
        for( i = 0; i < hullcount; i++ )
```

```
            {
                 Point pt = points[hull[i]];
                 line(img, pt0, pt,
                   Scalar(0, 255, 0), 1, CV_AA);
                 pt0 = pt;
            }
            imshow("hull", img);
            key = (char)waitKey();
            if( key == 27 || key == 'q' || key == 'Q' ) // 'ESC'
                break;
        }
        return 0;
}
```

执行结果如图 4-61 所示。

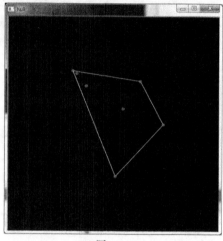

图 4-61

4.22 为轮廓建立许多矩形与圆形

为轮廓建立矩形与圆形的代码如下:

```
#include "opencv2/highgui/highgui.hpp"
#include "opencv2/imgproc/imgproc.hpp"
#include <iostream>
#include <stdio.h>
#include <stdlib.h>
using namespace cv;
using namespace std;
/// 声明全局变量
Mat src; Mat src_gray;
int thresh = 100;
int max_thresh = 255;
RNG rng(12345);
/// 声明函数
```

```cpp
void thresh_callback(int, void*);
int main(int, char** argv)
{
    /// 加载图形文件
    src = imread("C:\\images\\puzzle.png", 1);
    /// 转成灰度
    cvtColor(src, src_gray, COLOR_BGR2GRAY);
    /// 模糊
    blur(src_gray, src_gray, Size(3, 3));
    /// 建立窗口
    const char* source_window = "Source";
    namedWindow(source_window, WINDOW_AUTOSIZE);
    imshow(source_window, src);
    /// 建立滑杆
    createTrackbar(" Threshold:", "Source",
        &thresh, max_thresh, thresh_callback);
    thresh_callback(0, 0);
    waitKey(0);
    return(0);
}
void thresh_callback(int, void*)
{
    Mat threshold_output;
    vector<vector<Point> > contours;
    vector<Vec4i> hierarchy;
    /// 使用阈值检测边缘
    threshold(src_gray, threshold_output,
        thresh, 255, THRESH_BINARY);
    /// 寻找轮廓
    findContours(threshold_output, contours,
        hierarchy, RETR_TREE, CHAIN_APPROX_SIMPLE, Point(0, 0));
    /// 近似的多边形轮廓
    vector<vector<Point> > contours_poly(contours.size());
    /// 矩形
    vector<Rect> boundRect(contours.size());
    /// 圆形(圆心,半径)
    vector<Point2f>center(contours.size());
    vector<float>radius(contours.size());
    for (size_t i = 0; i < contours.size(); i++)
    {
        approxPolyDP(Mat(contours[i]), contours_poly[i], 3, true);
        boundRect[i] = boundingRect(Mat(contours_poly[i]));
        minEnclosingCircle(contours_poly[i], center[i], radius[i]);
    }
    /// 绘制多边形轮廓及许多矩形与圆形
    Mat drawing = Mat::zeros(threshold_output.size(), CV_8UC3);
    for (size_t i = 0; i< contours.size(); i++)
    {
        Scalar color = Scalar(rng.uniform(0, 255),
            rng.uniform(0, 255), rng.uniform(0, 255));
        /// 绘制多边形轮廓
        drawContours(drawing, contours_poly, (int)i,
            color, 1, 8, vector<Vec4i>(), 0, Point());
        /// 矩形
        rectangle(drawing, boundRect[i].tl(),
```

```
                boundRect[i].br(), color, 2, 8, 0);
        // 圆形
        circle(drawing, center[i], (int)radius[i], color, 2, 8, 0);
    }
    /// 显示结果
    namedWindow("Contours", WINDOW_AUTOSIZE);
    imshow("Contours", drawing);
}
```

程序说明

minEnclosingCircle(InputArray points, Point2f& center, float& radius)：寻找包围值定点最小区域的圆

（1）points：输入的二维向量点。

（2）center：输出圆的圆心。

（3）radius：输出圆的半径。

执行结果如图 4-62 所示。

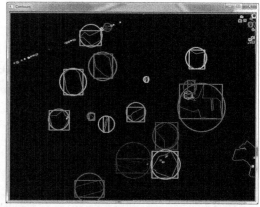

图 4-62

4.23 为轮廓建立旋转的矩形与椭圆形

为轮廓建立旋转的矩形与椭圆形的代码如下：

```
#include "opencv2/highgui/highgui.hpp"
#include "opencv2/imgproc/imgproc.hpp"
#include <iostream>
#include <stdio.h>
#include <stdlib.h>
using namespace cv;
using namespace std;
/// 声明全局变量
Mat src; Mat src_gray;
```

```cpp
int thresh = 100;
int max_thresh = 255;
RNG rng(12345);
/// 声明函数
void thresh_callback(int, void*);
int main(int, char** argv)
{
    /// 加载图形文件
    src = imread("C:\\images\\puzzle.png", 1);
    /// 变换成灰度
    cvtColor(src, src_gray, COLOR_BGR2GRAY);
    /// 模糊
    blur(src_gray, src_gray, Size(3, 3));
    /// 建立窗口
    const char* source_window = "Source";
    namedWindow(source_window, WINDOW_AUTOSIZE);
    imshow(source_window, src);
    /// 建立滑杆
    createTrackbar(" Threshold:", "Source",
        &thresh, max_thresh, thresh_callback);
    thresh_callback(0, 0);
    waitKey(0);
    return(0);
}
void thresh_callback(int, void*)
{
    Mat threshold_output;
    vector<vector<Point> > contours;
    vector<Vec4i> hierarchy;
    /// 使用阈值检测边缘
    threshold(src_gray, threshold_output,
        thresh, 255, THRESH_BINARY);
    /// 寻找轮廓
    findContours(threshold_output, contours, hierarchy,
        RETR_TREE, CHAIN_APPROX_SIMPLE, Point(0, 0));
    /// 为每个轮廓寻找旋转的矩形与椭圆形
    vector<RotatedRect> minRect(contours.size());
    vector<RotatedRect> minEllipse(contours.size());
    for (size_t i = 0; i < contours.size(); i++)
    {
        minRect[i] = minAreaRect(Mat(contours[i]));
        if (contours[i].size() > 5)
        {
            minEllipse[i] = fitEllipse(Mat(contours[i]));
        }
    }
    /// 绘制轮廓、旋转的矩形、椭圆形
    Mat drawing = Mat::zeros(threshold_output.size(), CV_8UC3);
    for (size_t i = 0; i< contours.size(); i++)
    {
        Scalar color = Scalar(rng.uniform(0, 255),
            rng.uniform(0, 255), rng.uniform(0, 255));
        // 轮廓
        drawContours(drawing, contours, (int)i, color,
```

```
            1, 8, vector<Vec4i>(), 0, Point());
        // 椭圆形
        ellipse(drawing, minEllipse[i], color, 2, 8);
        // 旋转的矩形
        Point2f rect_points[4]; minRect[i].points(rect_points);
        for (int j = 0; j < 4; j++)
            line(drawing, rect_points[j],
                rect_points[(j + 1) % 4], color, 1, 8);
    }
    /// 显示结果
    namedWindow("Contours", WINDOW_AUTOSIZE);
    imshow("Contours", drawing);
}
```

程序说明

RotatedRect minAreaRect(InputArray points)：寻找包围指定点最小区域的旋转矩形

points：输入的二维向量点。

执行结果如图 4-63 所示。

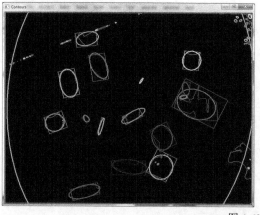

图 4-63

4.24 图像矩

图像矩的代码如下：

```
#include "opencv2/highgui/highgui.hpp"
#include "opencv2/imgproc/imgproc.hpp"
#include <iostream>
#include <stdio.h>
#include <stdlib.h>
using namespace cv;
using namespace std;
```

```cpp
/// 声明全局变量
Mat src; Mat src_gray;
int thresh = 100;
int max_thresh = 255;
RNG rng(12345);
/// 声明函数
void thresh_callback(int, void*);
int main(int, char** argv)
{
    /// 加载图形文件
    src = imread("C:\\images\\board.jpg", 1);
    /// 变换成灰度
    cvtColor(src, src_gray, COLOR_BGR2GRAY);
    /// 模糊
    blur(src_gray, src_gray, Size(3, 3));
    /// 建立窗口
    const char* source_window = "Source";
    namedWindow(source_window, WINDOW_AUTOSIZE);
    imshow(source_window, src);
    /// 建立滑杆
    createTrackbar(" Canny thresh:", "Source",
        &thresh, max_thresh, thresh_callback);
    thresh_callback(0, 0);
    waitKey(0);
    return(0);
}
void thresh_callback(int, void*)
{
    Mat canny_output;
    vector<vector<Point> > contours;
    vector<Vec4i> hierarchy;
    /// 使用canny 检测边缘
    Canny(src_gray, canny_output, thresh, thresh * 2, 3);
    /// Find contours
    findContours(canny_output, contours, hierarchy,
        RETR_TREE, CHAIN_APPROX_SIMPLE, Point(0, 0));
    /// 取得矩(moments)
    vector<Moments> mu(contours.size());
    for (size_t i = 0; i < contours.size(); i++)
    {
        mu[i] = moments(contours[i], false);
    }
    ///  Get the mass centers:
    vector<Point2f> mc(contours.size());
    for (size_t i = 0; i < contours.size(); i++)
    {
        mc[i] = Point2f(static_cast<float>(mu[i].m10 / mu[i].m00),
            static_cast<float>(mu[i].m01 / mu[i].m00));
    }
    /// 绘制轮廓
    Mat drawing = Mat::zeros(canny_output.size(), CV_8UC3);
    for (size_t i = 0; i< contours.size(); i++)
    {
```

```
        Scalar color = Scalar(rng.uniform(0, 255),
            rng.uniform(0, 255), rng.uniform(0, 255));
        drawContours(drawing, contours, (int)i, color
            , 2, 8, hierarchy, 0, Point());
        circle(drawing, mc[i], 4, color, -1, 8, 0);
    }
    /// 显示结果
    namedWindow("Contours", WINDOW_AUTOSIZE);
    imshow("Contours", drawing);
    /// 用矩(moments) 00 计算区域并与结果比较
    printf("\t Info: Area and Contour Length \n");
    for (size_t i = 0; i< contours.size(); i++)
    {
        printf(" * Contour[%d] - Area (M_00) = %.2f - Area OpenCV: %.2f -
            Length: %.2f \n",
            (int)i, mu[i].m00, contourArea(contours[i]),
            arcLength(contours[i], true));
        Scalar color = Scalar(rng.uniform(0, 255),
            rng.uniform(0, 255), rng.uniform(0, 255));
        drawContours(drawing, contours, (int)i, color,
            2, 8, hierarchy, 0, Point());
        circle(drawing, mc[i], 4, color, -1, 8, 0);
    }
}
```

程序说明

Moments moments(InputArray array, bool binaryImage=false)：计算第三级多边形或点阵化的形状（rasterized shape）的矩（moments）

（1）array：点阵化的图像。

（2）binaryImage：如果值为 true，图像所有非 0 的像素都视为 1。

执行结果如图 4-64 所示。

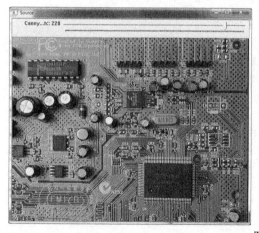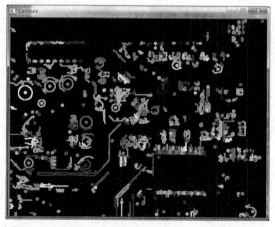

图 4-64

4.25 点多边形测试

点多边形测试的代码如下:

```cpp
#include "opencv2/highgui/highgui.hpp"
#include "opencv2/imgproc/imgproc.hpp"
#include <iostream>
#include <stdio.h>
#include <stdlib.h>
using namespace cv;
using namespace std;
int main(void)
{
    /// 建立图像
    const int r = 100;
    Mat src = Mat::zeros(Size(4 * r, 4 * r), CV_8UC1);
    /// 用循序的点建立轮廓
    vector<Point2f> vert(6);
    vert[0] = Point(3 * r / 2, static_cast<int>(1.34*r));
    vert[1] = Point(1 * r, 2 * r);
    vert[2] = Point(3 * r / 2, static_cast<int>(2.866*r));
    vert[3] = Point(5 * r / 2, static_cast<int>(2.866*r));
    vert[4] = Point(3 * r, 2 * r);
    vert[5] = Point(5 * r / 2, static_cast<int>(1.34*r));
    /// Draw it in src
    for (int j = 0; j < 6; j++)
    {
        line(src, vert[j], vert[(j + 1) % 6], Scalar(255), 3, 8);
    }
    /// 取得轮廓
    vector<vector<Point> > contours; vector<Vec4i> hierarchy;
    Mat src_copy = src.clone();
    findContours(src_copy, contours, hierarchy,
        RETR_TREE, CHAIN_APPROX_SIMPLE);
    /// 计算轮廓距离
    Mat raw_dist(src.size(), CV_32FC1);
    for (int j = 0; j < src.rows; j++)
    {
        for (int i = 0; i < src.cols; i++)
        {
            raw_dist.at<float>(j, i)
                = (float)pointPolygonTest(contours[0],
                    Point2f((float)i, (float)j), true);
        }
    }
    double minVal; double maxVal;
    minMaxLoc(raw_dist, &minVal, &maxVal, 0, 0, Mat());
    minVal = abs(minVal); maxVal = abs(maxVal);
    /// 描绘图形上的距离
    Mat drawing = Mat::zeros(src.size(), CV_8UC3);
    for (int j = 0; j < src.rows; j++)
```

```
        {
            for (int i = 0; i < src.cols; i++)
            {
                if (raw_dist.at<float>(j, i) < 0)
                {
                    drawing.at<Vec3b>(j, i)[0] =
                            (uchar)(255 - abs(raw_dist.at<float>(j, i))
                            * 255 / minVal);
                }
                else if (raw_dist.at<float>(j, i) > 0)
                {
                    drawing.at<Vec3b>(j, i)[2] =
                            (uchar)(255 - raw_dist.at<float>(j, i)
                            * 255 / maxVal);
                }
                else
                {
                    drawing.at<Vec3b>(j, i)[0] = 255;
                    drawing.at<Vec3b>(j, i)[1] = 255;
                    drawing.at<Vec3b>(j, i)[2] = 255;
                }
            }
        }
        /// 建立窗口并显示结果
        const char* source_window = "Source";
        namedWindow(source_window, WINDOW_AUTOSIZE);
        imshow(source_window, src);
        namedWindow("Distance", WINDOW_AUTOSIZE);
        imshow("Distance", drawing);
        waitKey(0);
        return(0);
}
```

执行结果如图 4-65 所示。

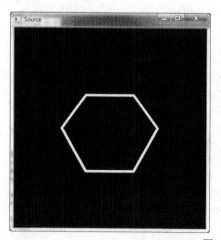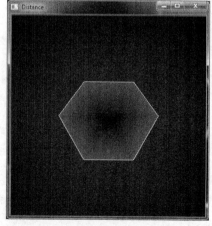

图 4-65

4.26 线性变换

线性变换的代码如下：

```cpp
#include "opencv2/highgui/highgui.hpp"
using namespace cv;
/** 声明全局变量 */
const int alpha_max = 5;
const int beta_max = 125;
int alpha; /** 简单对比度控制 */
int beta;  /** 简单亮度控制 */
/** 图像数据 */
Mat image;
/**
 * 对比度与亮度值改变处理的函数
 */
static void on_trackbar(int, void*)
{
   Mat new_image = Mat::zeros(image.size(), image.type());
   for (int y = 0; y < image.rows; y++)
      for (int x = 0; x < image.cols; x++)
         for (int c = 0; c < 3; c++)
            new_image.at<Vec3b>(y, x)[c] =
               saturate_cast<uchar>(alpha*(image.at<Vec3b>(y, x)[c]) + beta);
   imshow("New Image", new_image);
}
int main(int, char** argv)
{
   /// 加载图形文件
   image = imread("C:\\images\\lena.jpg");
   /// 起始变量值
   alpha = 1;
   beta = 0;
   /// 建立窗口
   namedWindow("Original Image", 1);
   namedWindow("New Image", 1);
   /// 建立滑杆
   createTrackbar("Contrast Trackbar", "New Image",
      &alpha, alpha_max, on_trackbar);
   createTrackbar("Brightness Trackbar", "New Image",
      &beta, beta_max, on_trackbar);
   /// 显示结果
   imshow("Original Image", image);
   imshow("New Image", image);
   /// 等待按键
   waitKey();
   return 0;
}
```

本程序是通过线性变换来改变亮度和对比度，如图 4-66 所示。

（a）原图

（b）改变亮度

（c）改变亮度与对比度

图 4-66

第 5 章 Calib3d 模块

虽然图像都是二维平面的，但摄源都源自于三维世界。相机校准（camera calibration）的原理是修正单眼通过镜头造成的偏差，所以相机校准对对象的感知，特别是三维空间的测量十分重要。通过相机校准取得几何与偏差模型，我们可以将实体世界通过像素而重建三维的感知环境。

要校准是因为光学偏差，而光学偏差属于光学理论，如果读者想更进一步获得光学相关知识，可参考如下链接提供的免费教学资料 https://www.coursera.org/learn/ji-chu-guang-xue。

针对每台相机校准只需执行一次，然后将成功校准的参数存储成 XML 或 YAML 格式，以后再使用时就可以使用程序载入校准存储的参数，存储方式请参考下例程序中 runCalibrationAndSave 函数。

5.1 使用棋盘进行相机校准

使用棋盘进行相机校准的代码如下：

```cpp
#include <iostream>
#include <sstream>
#include <time.h>
#include <stdio.h>
#include <opencv2/core/core.hpp>
#include <opencv2/imgproc/imgproc.hpp>
#include <opencv2/calib3d/calib3d.hpp>
#include <opencv2/highgui/highgui.hpp>
#ifndef _CRT_SECURE_NO_WARNINGS
# define _CRT_SECURE_NO_WARNINGS
#endif
using namespace cv;
using namespace std;
class Settings
{
public:
    Settings() : goodInput(false) {}
    enum Pattern { NOT_EXISTING, CHESSBOARD, CIRCLES_GRID,
                   ASYMMETRIC_CIRCLES_GRID };
    enum InputType { INVALID, CAMERA, VIDEO_FILE, IMAGE_LIST };
    // 此类的输出功能
    void write(FileStorage& fs) const
    {
        fs << "{" << "BoardSize_Width" << boardSize.width
            << "BoardSize_Height" << boardSize.height
            << "Square_Size" << squareSize
            << "Calibrate_Pattern" << patternToUse
            << "Calibrate_NrOfFrameToUse" << nrFrames
            << "Calibrate_FixAspectRatio" << aspectRatio
            << "Calibrate_AssumeZeroTangentialDistortion"
            << calibZeroTangentDist
            << "Calibrate_FixPrincipalPointAtTheCenter"
            << calibFixPrincipalPoint
```

```cpp
            << "Write_DetectedFeaturePoints" << bwritePoints
            << "Write_extrinsicParameters" << bwriteExtrinsics
            << "Write_outputFileName" << outputFileName
            << "Show_UndistortedImage" << showUndistorsed
            << "Input_FlipAroundHorizontalAxis" << flipVertical
            << "Input_Delay" << delay
            << "Input" << input
            << "}";
    }

    // 此类的读取功能
    void read(const FileNode& node)
    {
        node["BoardSize_Width"] >> boardSize.width;
        node["BoardSize_Height"] >> boardSize.height;
        node["Calibrate_Pattern"] >> patternToUse;
        node["Square_Size"] >> squareSize;
        node["Calibrate_NrOfFrameToUse"] >> nrFrames;
        node["Calibrate_FixAspectRatio"] >> aspectRatio;
        node["Write_DetectedFeaturePoints"] >> bwritePoints;
        node["Write_extrinsicParameters"] >> bwriteExtrinsics;
        node["Write_outputFileName"] >> outputFileName;
        node["Calibrate_AssumeZeroTangentialDistortion"]
            >> calibZeroTangentDist;
        node["Calibrate_FixPrincipalPointAtTheCenter"]
            >> calibFixPrincipalPoint;
        node["Input_FlipAroundHorizontalAxis"] >> flipVertical;
        node["Show_UndistortedImage"] >> showUndistorsed;
        node["Input"] >> input;
        node["Input_Delay"] >> delay;
        interprate();
    }
    // 解析棋盘
    void interprate()
    {
        goodInput = true;
        if (boardSize.width <= 0 || boardSize.height <= 0)
        {
            cerr << "棋盘大小无效: "
                << boardSize.width << " "
                << boardSize.height << endl;
            goodInput = false;
        }
        if (squareSize <= 10e-6)
        {
            cerr << "棋盘方块大小无效 " << squareSize << endl;
            goodInput = false;
        }
        if (nrFrames <= 0)
        {
            cerr << "帧数无效 " << nrFrames << endl;
            goodInput = false;
        }
        // 检查有效的输入
        if (input.empty())
            inputType = INVALID;
```

```cpp
        else
        {
            if (input[0] >= '0' && input[0] <= '9')
            {
                stringstream ss(input);
                ss >> cameraID;
                inputType = CAMERA;
            }
            else
            {
                if (readStringList(input, imageList))
                {
                 inputType = IMAGE_LIST;
                 nrFrames = (nrFrames <
                     (int)imageList.size()) ? nrFrames : (int)imageList.size();
                }
                else
                  inputType = VIDEO_FILE;
            }
            if (inputType == CAMERA)
                inputCapture.open(cameraID);
            if (inputType == VIDEO_FILE)
                inputCapture.open(input);
            if (inputType != IMAGE_LIST && !inputCapture.isOpened())
                inputType = INVALID;
        }
        if (inputType == INVALID)
        {
            cerr << " 不存在的输入: " << input;
            goodInput = false;
        }
        flag = 0;
        if (calibFixPrincipalPoint)
            flag |= CV_CALIB_FIX_PRINCIPAL_POINT;
        if (calibZeroTangentDist)
            flag |= CV_CALIB_ZERO_TANGENT_DIST;
        if (aspectRatio)
            flag |= CV_CALIB_FIX_ASPECT_RATIO;
        calibrationPattern = NOT_EXISTING;
        if (!patternToUse.compare("CHESSBOARD"))
            calibrationPattern = CHESSBOARD;
        if (!patternToUse.compare("CIRCLES_GRID"))
            calibrationPattern = CIRCLES_GRID;
        if (!patternToUse.compare("ASYMMETRIC_CIRCLES_GRID"))
            calibrationPattern = ASYMMETRIC_CIRCLES_GRID;
        if (calibrationPattern == NOT_EXISTING)
        {
            cerr << " Inexistent camera calibration mode: "
                << patternToUse << endl;
            goodInput = false;
        }
        atImageList = 0;
    }
    // 读取下一个图像文件
    Mat nextImage()
```

```cpp
    {
        Mat result;
        if (inputCapture.isOpened())
        {
            Mat view0;
            inputCapture >> view0;
            view0.copyTo(result);
        }
        else if (atImageList < (int)imageList.size())
            result = imread(imageList[atImageList++],
                            CV_LOAD_IMAGE_COLOR);
        return result;
    }
    static bool readStringList(const string& filename,
                               vector<string>& l)
    {
        l.clear();
        FileStorage fs(filename, FileStorage::READ);
        if (!fs.isOpened())
            return false;
        FileNode n = fs.getFirstTopLevelNode();
        if (n.type() != FileNode::SEQ)
            return false;
        FileNodeIterator it = n.begin(), it_end = n.end();
        for (; it != it_end; ++it)
            l.push_back((string)*it);
        return true;
    }
public:
    // 棋盘大小
    Size boardSize;

    // One of the Chessboard, circles, or asymmetric circle pattern
    Pattern calibrationPattern;

    // 棋盘内方块大小
    float squareSize;

    // 用来校准所输入的帧数
    int nrFrames;

    // 外观比例(aspect ratio)
    float aspectRatio;

    // 拍摄输入的暂停时间
    int delay;

    //  输出检测的特侦点
    bool bwritePoints;

    // 输出外在的(extrinsic)参数
    bool bwriteExtrinsics;

    // 零切向畸变(zero tangential distortion)
    bool calibZeroTangentDist;
```

```cpp
        // 修正中心的主要点(principal point)
        bool calibFixPrincipalPoint;

        // 在水平坐标翻转抓到的图像
        bool flipVertical;

        // 输出的文档名
        string outputFileName;

        // 校准后显示失真图像
        bool showUndistorsed;

        // 输入的文档名
        string input;
        int cameraID;
        vector<string> imageList;
        int atImageList;
        VideoCapture inputCapture;
        InputType inputType;
        bool goodInput;
        int flag;
private:
        string patternToUse;
};
static void read(const FileNode& node, Settings& x,
    const Settings& default_value = Settings())
{
    if (node.empty())
        x = default_value;
    else
        x.read(node);
}
enum { DETECTION = 0, CAPTURING = 1, CALIBRATED = 2 };
// 存储校准参数
bool runCalibrationAndSave(Settings& s, Size imageSize,
    Mat& cameraMatrix, Mat& distCoeffs,
    vector<vector<Point2f> > imagePoints);
int main(int argc, char* argv[])
{
    Settings s;
    const string inputSettingsFile
        = argc > 1 ? argv[1] : "C:\\images\\default.xml";

    // 读取设置(default.xml)
    FileStorage fs(inputSettingsFile, FileStorage::READ);
    if (!fs.isOpened())
    {
        cout << "无法打开设置配置文件: \""
            << inputSettingsFile << "\"" << endl;
        getchar();
        return -1;
    }
    fs["Settings"] >> s;
    // 关闭配置文件
```

```cpp
    fs.release();
    if (!s.goodInput)
    {
        cout << "输入的检测文件无效. 程序结束. " << endl;
        getchar();
        return -1;
    }
    vector<vector<Point2f> > imagePoints;
    Mat cameraMatrix, distCoeffs;
    Size imageSize;
    int mode = s.inputType ==
        Settings::IMAGE_LIST ? CAPTURING : DETECTION;
    clock_t prevTimestamp = 0;
    const Scalar RED(0, 0, 255), GREEN(0, 255, 0);
    const char ESC_KEY = 27;
    // 图像文件读完自动结束，不用设 for loop 结束条件
    for (int i = 0;; ++i)
    {
        Mat view;
        bool blinkOutput = false;
        // 读取下一个图像文件
        view = s.nextImage();
        //-----  如果没有图像或读取完毕就结束校准并显示结果 -----
        if (mode == CAPTURING && imagePoints.size()
            >= (unsigned)s.nrFrames)
        {
            if (runCalibrationAndSave(s, imageSize,
                cameraMatrix, distCoeffs, imagePoints))
                mode = CALIBRATED;
            else
                mode = DETECTION;
        }

        // 如果没有图像就开始校准,存储并结束循环
        if (view.empty())
        {
            if (imagePoints.size() > 0)
                // 因为 imagePoints 是空的所以都未为存储
                runCalibrationAndSave(s, imageSize,
                    cameraMatrix, distCoeffs, imagePoints);
            break;
        }
        // 相机须要翻转图像，配置文件中未定义要垂直翻转
        // 如果是手机拍照一般是需要左右翻转图像
        imageSize = view.size();
        if (s.flipVertical)
            flip(view, view, 0);
        vector<Point2f> pointBuf;
        bool found;
        // 寻找目前输入文件的样式(pattern)
        switch (s.calibrationPattern)
        {
        // 配置文件只定义此样式
        case Settings::CHESSBOARD:
            found = findChessboardCorners(view,
```

```cpp
                s.boardSize, pointBuf, CV_CALIB_CB_ADAPTIVE_THRESH |
                CV_CALIB_CB_FAST_CHECK | CV_CALIB_CB_NORMALIZE_IMAGE);
            break;
        case Settings::CIRCLES_GRID:
            found = findCirclesGrid(view, s.boardSize, pointBuf);
            break;
        case Settings::ASYMMETRIC_CIRCLES_GRID:
            found = findCirclesGrid(view, s.boardSize, pointBuf,
                CALIB_CB_ASYMMETRIC_GRID);
            break;
        default:
            found = false;
            break;
        }
        // 找到样式
        if (found)
        {
            // 改善找到的棋盘角坐标
            if (s.calibrationPattern == Settings::CHESSBOARD)
            {
                Mat viewGray;
                cvtColor(view, viewGray, COLOR_BGR2GRAY);
                cornerSubPix(viewGray, pointBuf,
                    Size(11, 11), Size(-1, -1),
                    TermCriteria(CV_TERMCRIT_EPS + CV_TERMCRIT_ITER, 30, 0.1));
            }
            // 拍摄方式
            if (mode == CAPTURING &&
                (!s.inputCapture.isOpened() || clock() -
                    prevTimestamp > s.delay*1e-3*CLOCKS_PER_SEC))
            {
                imagePoints.push_back(pointBuf);
                prevTimestamp = clock();
                blinkOutput = s.inputCapture.isOpened();
            }
            // 绘制角
            drawChessboardCorners(view, s.boardSize,
                Mat(pointBuf), found);
        }
        //--------------- 文字输出 ---------------------
        string msg = (mode == CAPTURING) ? "100/100" :
            mode == CALIBRATED ? "Calibrated" : "Press 'g' to start";
        int baseLine = 0;
        Size textSize = getTextSize(msg, 1, 1, 1, &baseLine);
        Point textOrigin(view.cols - 2 * textSize.width - 10,
            view.rows - 2 * baseLine - 10);
        if (mode == CAPTURING)
        {
            // 非畸变，作者加入 i 是要查看确认是第几个图
            // 原程序有问题，imagePoints 是空的
            if (s.showUndistorsed)
                msg = format("%d: %d/%d Undist", i,
                    (int)imagePoints.size(), s.nrFrames);
            else
                msg = format("%d: %d/%d", i, (int)imagePoints.size(),
```

```cpp
                        s.nrFrames);
        putText(view, msg, textOrigin, 1, 1,
                mode == CALIBRATED ? GREEN : RED);
        if (blinkOutput)
            bitwise_not(view, view);
        // 如果要拍摄时则要校准并取得相机畸变系数(distortion coefficients)
        if (mode == CALIBRATED && s.showUndistorsed)
        {
            Mat temp = view.clone();
            // 修正图像
            undistort(temp, view, cameraMatrix, distCoeffs);
        }
        //------------------ 显示图像并检查输入指令 -----------------
        imshow("Image View", view);
        char key = (char)waitKey(s.inputCapture.isOpened() ? 50 : s.delay);
        // 结束循环
        if (key == ESC_KEY)
            break;
        // 是否显示未畸变
        if (key == 'u' && mode == CALIBRATED)
            s.showUndistorsed = !s.showUndistorsed;
        // 重新检测
        if (s.inputCapture.isOpened() && key == 'g')
        {
            mode = CAPTURING;
            imagePoints.clear();
        }
    }
    // -----------------显示在图列中未畸变的图像------------------
    if (s.inputType == Settings::IMAGE_LIST && s.showUndistorsed)
    {
        Mat view, rview, map1, map2;
        initUndistortRectifyMap(cameraMatrix, distCoeffs, Mat(),
            getOptimalNewCameraMatrix(cameraMatrix, distCoeffs,
            imageSize, 1, imageSize, 0),
            imageSize, CV_16SC2, map1, map2);
        for (int i = 0; i < (int)s.imageList.size(); i++)
        {
            view = imread(s.imageList[i], 1);
            if (view.empty())
                continue;
            remap(view, rview, map1, map2, INTER_LINEAR);
            imshow("Image View", rview);
            char c = (char)waitKey();
            if (c == ESC_KEY || c == 'q' || c == 'Q')
                break;
        }
    }
    return 0;
}
static double computeReprojectionErrors(
    const vector<vector<Point3f> >& objectPoints,
    const vector<vector<Point2f> >& imagePoints,
    const vector<Mat>& rvecs, const vector<Mat>& tvecs,
```

```cpp
        const Mat& cameraMatrix, const Mat& distCoeffs,
        vector<float>& perViewErrors)
{
    vector<Point2f> imagePoints2;
    int i, totalPoints = 0;
    double totalErr = 0, err;
    perViewErrors.resize(objectPoints.size());
    for (i = 0; i < (int)objectPoints.size(); ++i)
    {
        projectPoints(Mat(objectPoints[i]), rvecs[i],
            tvecs[i], cameraMatrix,
            distCoeffs, imagePoints2);
        err = norm(Mat(imagePoints[i]), Mat(imagePoints2), CV_L2);
        int n = (int)objectPoints[i].size();
        perViewErrors[i] = (float)std::sqrt(err*err / n);
        totalErr += err*err;
        totalPoints += n;
    }
    return std::sqrt(totalErr / totalPoints);
}
static void calcBoardCornerPositions(Size boardSize,
    float squareSize, vector<Point3f>& corners,
    Settings::Pattern patternType)
{
    corners.clear();
    switch (patternType)
    {
    case Settings::CHESSBOARD:
    case Settings::CIRCLES_GRID:
        for (int i = 0; i < boardSize.height; ++i)
            for (int j = 0; j < boardSize.width; ++j)
                corners.push_back(Point3f(float(j*squareSize),
                    float(i*squareSize), 0));
        break;
    case Settings::ASYMMETRIC_CIRCLES_GRID:
        for (int i = 0; i < boardSize.height; i++)
            for (int j = 0; j < boardSize.width; j++)
                corners.push_back(Point3f(float((2 * j + i % 2)*squareSize),
                    float(i*squareSize), 0));
        break;
    default:
        break;
    }
}
static bool runCalibration(Settings& s, Size& imageSize,
    Mat& cameraMatrix, Mat& distCoeffs,
    vector<vector<Point2f> > imagePoints,
    vector<Mat>& rvecs, vector<Mat>& tvecs,
    vector<float>& reprojErrs, double& totalAvgErr)
{
    cameraMatrix = Mat::eye(3, 3, CV_64F);
    if (s.flag & CV_CALIB_FIX_ASPECT_RATIO)
        cameraMatrix.at<double>(0, 0) = 1.0;
    distCoeffs = Mat::zeros(8, 1, CV_64F);
    vector<vector<Point3f> > objectPoints(1);
```

```cpp
        calcBoardCornerPositions(s.boardSize, s.squareSize,
            objectPoints[0], s.calibrationPattern);
        objectPoints.resize(imagePoints.size(), objectPoints[0]);
        // 寻找相机固有的(intrinsic)与外在的(extrinsic) 参数
        double rms = calibrateCamera(objectPoints, imagePoints,
            imageSize, cameraMatrix, distCoeffs, rvecs,
            tvecs, s.flag | CV_CALIB_FIX_K4 | CV_CALIB_FIX_K5);
        cout << "Re-projection error reported by calibrateCamera: "
            << rms << endl;
        bool ok = checkRange(cameraMatrix)
            && checkRange(distCoeffs);
        totalAvgErr = computeReprojectionErrors(objectPoints,
            imagePoints, rvecs, tvecs, cameraMatrix,
            distCoeffs, reprojErrs);
        return ok;
}
// 打印相机参数到输出文件
static void saveCameraParams(Settings& s, Size& imageSize,
    Mat& cameraMatrix, Mat& distCoeffs,
    const vector<Mat>& rvecs, const vector<Mat>& tvecs,
    const vector<float>& reprojErrs,
    const vector<vector<Point2f> >& imagePoints,
    double totalAvgErr)
{
    FileStorage fs(s.outputFileName, FileStorage::WRITE);
    time_t tm;
    time(&tm);
    struct tm *t2 = localtime(&tm);
    char buf[1024];
    strftime(buf, sizeof(buf) - 1, "%c", t2);
    fs << "calibration_Time" << buf;
    if (!rvecs.empty() || !reprojErrs.empty())
        fs << "nrOfFrames"
            << (int)std::max(rvecs.size(), reprojErrs.size());
    fs << "image_Width" << imageSize.width;
    fs << "image_Height" << imageSize.height;
    fs << "board_Width" << s.boardSize.width;
    fs << "board_Height" << s.boardSize.height;
    fs << "square_Size" << s.squareSize;
    if (s.flag & CV_CALIB_FIX_ASPECT_RATIO)
        fs << "FixAspectRatio" << s.aspectRatio;
    if (s.flag)
    {
        sprintf(buf, "flags: %s%s%s%s",
            s.flag & CV_CALIB_USE_INTRINSIC_GUESS ? " +use_intrinsic_guess" : "",
            s.flag & CV_CALIB_FIX_ASPECT_RATIO ? " +fix_aspectRatio" : "",
            s.flag & CV_CALIB_FIX_PRINCIPAL_POINT ? " +fix_principal_point" : "",
            s.flag & CV_CALIB_ZERO_TANGENT_DIST ? " +zero_tangent_dist" : "");
        cvWriteComment(*fs, buf, 0);
    }
    fs << "flagValue" << s.flag;
    fs << "Camera_Matrix" << cameraMatrix;
    fs << "Distortion_Coefficients" << distCoeffs;
    fs << "Avg_Reprojection_Error" << totalAvgErr;
    if (!reprojErrs.empty())
```

```cpp
            fs << "Per_View_Reprojection_Errors" << Mat(reprojErrs);
        if (!rvecs.empty() && !tvecs.empty())
        {
            CV_Assert(rvecs[0].type() == tvecs[0].type());
            Mat bigmat((int)rvecs.size(), 6, rvecs[0].type());
            for (int i = 0; i < (int)rvecs.size(); i++)
            {
                Mat r = bigmat(Range(i, i + 1), Range(0, 3));
                Mat t = bigmat(Range(i, i + 1), Range(3, 6));
                CV_Assert(rvecs[i].rows == 3 && rvecs[i].cols == 1);
                CV_Assert(tvecs[i].rows == 3 && tvecs[i].cols == 1);
                r = rvecs[i].t();
                t = tvecs[i].t();
            }
            cvWriteComment(*fs, "a set of 6-tuples (rotation vector + 
                    translation vector) for each view", 0);
            fs << "Extrinsic_Parameters" << bigmat;
        }
        if (!imagePoints.empty())
        {
            Mat imagePtMat((int)imagePoints.size(),
                            (int)imagePoints[0].size(), CV_32FC2);
            for (int i = 0; i < (int)imagePoints.size(); i++)
            {
                Mat r = imagePtMat.row(i).reshape(2, imagePtMat.cols);
                Mat imgpti(imagePoints[i]);
                imgpti.copyTo(r);
            }
            fs << "Image_points" << imagePtMat;
        }
}
// 存储校准参数
bool runCalibrationAndSave(Settings& s, Size imageSize,
    Mat& cameraMatrix, Mat& distCoeffs,
    vector<vector<Point2f> > imagePoints)
{
    vector<Mat> rvecs, tvecs;
    vector<float> reprojErrs;
    double totalAvgErr = 0;
    bool ok = runCalibration(s, imageSize, cameraMatrix,
        distCoeffs, imagePoints, rvecs, tvecs,
        reprojErrs, totalAvgErr);
    cout << (ok ? "校准成功" : "校准失败")
        << ". avg re projection error = " << totalAvgErr;
    if (ok)
        saveCameraParams(s, imageSize, cameraMatrix,
            distCoeffs, rvecs, tvecs, reprojErrs,
            imagePoints, totalAvgErr);
    return ok;
}
```

要观看本程序的实际执行过程，请参考如下链接 https://www.youtube.com/watch?v=ViPN810E0SU，如图 5-1 所示。

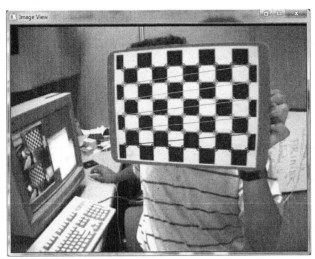

图 5-1

程序中配置文件 default.xml 的内容,主要是校准的设置,因为程序是根据设置来决定程序执行,所以可以修改此配置文件以了解程序各功能。配置文件 default.xml 的内容如下:

```
<?xml version="1.0"?>
<opencv_storage>
<Settings>
<!-- 每个项目行与列的内部角数(Number of inner corners). -->
<BoardSize_Width> 9</BoardSize_Width>
<BoardSize_Height>6</BoardSize_Height>
<!-- 方格大小使用用户定义的单位(像素,毫米)-->
<Square_Size>50</Square_Size>
<!-- 校准输入形态: CHESSBOARD CIRCLES_GRID ASYMMETRIC_CIRCLES_GRID -->
<Calibrate_Pattern>"CHESSBOARD"</Calibrate_Pattern>
<!-- 校准的输入.
相机 -> 用相机代号,例如 "1"
视频(video)   -> 图像文件路径,例如 "/tmp/x.avi"
图像清单(image list)  -> XML 或 YAML 文件路径并含有图像文件,例如 "/tmp/circles_list.xml"
-->
<Input>"c:/images/VID5.xml"</Input>
<!-- 如果是 true (非零值) 水平翻转输入图像.-->
<Input_FlipAroundHorizontalAxis>0</Input_FlipAroundHorizontalAxis>
<!-- 拍摄时等待时间 -->
<Input_Delay>100</Input_Delay>
<!-- 使用几个帧进行校准 -->
<Calibrate_NrOfFrameToUse>25</Calibrate_NrOfFrameToUse>
<!-- 只考虑 fy 是自由参数(free parameter), fx/fy 比设为与输入的 cameraMatrix 相
同. 0 - 不设置,非零值 - 使用 -->
<Calibrate_FixAspectRatio> 1 </Calibrate_FixAspectRatio>
<!-- 如果 true (非零值) 正切畸变系数(tangential distortion coefficients)设为零
并保持为零 -->
<Calibrate_AssumeZeroTangentialDistortion>1</Calibrate_AssumeZeroTangentialDistortion>
<!-- 如果 true (非零值)全域优化时主点(principal point)不改变-->
```

```
<Calibrate_FixPrincipalPointAtTheCenter> 1 </Calibrate_
FixPrincipalPointAtTheCenter>
<!-- 输出文档名 -->
<Write_outputFileName>" C:/images/process/out_camera_data.xml"</Write_
outputFileName>
<!-- 如果 true (非零值) 输出特征点-->
<Write_DetectedFeaturePoints>1</Write_DetectedFeaturePoints>
<!-- 如果 true (非零值)输出相机固有的参数(extrinsic parameters.-->
<Write_extrinsicParameters>1</Write_extrinsicParameters>
<!-- 如果 true (非零值) 校准后显示非畸变图像 -->
<Show_UndistortedImage>1</Show_UndistortedImage>
</Settings>
</opencv_storage>
```

default.xml 文件中 VID5.xml 的内容, 实际图像文件位置, 代码如下:

```
<?xml version="1.0"?>
<opencv_storage>
<imagelist>
"c:/images/Chessboards/left01.jpg"
"c:/images/Chessboards/right01.jpg"
"c:/images/Chessboards/left02.jpg"
"c:/images/Chessboards/right02.jpg"
"c:/images/Chessboards/left03.jpg"
"c:/images/Chessboards/right03.jpg"
"c:/images/Chessboards/left04.jpg"
"c:/images/Chessboards/right04.jpg"
"c:/images/Chessboards/left05.jpg"
"c:/images/Chessboards/right05.jpg"
"c:/images/Chessboards/left06.jpg"
"c:/images/Chessboards/right06.jpg"
"c:/images/Chessboards/left07.jpg"
"c:/images/Chessboards/right07.jpg"
"c:/images/Chessboards/left08.jpg"
"c:/images/Chessboards/right08.jpg"
"c:/images/Chessboards/left09.jpg"
"c:/images/Chessboards/right09.jpg"
"c:/images/Chessboards/left11.jpg"
"c:/images/Chessboards/right11.jpg"
"c:/images/Chessboards/left12.jpg"
"c:/images/Chessboards/right12.jpg"
"c:/images/Chessboards/left13.jpg"
"c:/images/Chessboards/right13.jpg"
"c:/images/Chessboards/left14.jpg"
"c:/images/Chessboards/right14.jpg"
</imagelist>
</opencv_storage>
```

自我校准

自我校准的代码如下:

```
#include "opencv2/calib3d/calib3d.hpp"
#include "opencv2/imgproc/imgproc.hpp"
#include "opencv2/highgui/highgui.hpp"
#include <iostream>
```

```cpp
#include <vector>
#include <algorithm>
#include <iterator>
#include <stdio.h>
using namespace cv;
using namespace std;
// 将下列类声明为与 OpenCV 同命名空间
namespace cv
{
class ChessBoardGenerator
{
public:
    double sensorWidth;
    double sensorHeight;
    size_t squareEdgePointsNum;
    double min_cos;
    mutable double cov;
    Size patternSize;
    int rendererResolutionMultiplier;
    ChessBoardGenerator(const Size& patternSize = Size(8, 6));
    Mat operator()(const Mat& bg, const Mat& camMat,
        const Mat& distCoeffs, vector<Point2f>& corners) const;
    Size cornersSize() const;
private:
    void generateEdge(const Point3f& p1, const Point3f& p2,
        vector<Point3f>& out) const;
    Mat generageChessBoard(const Mat& bg, const Mat& camMat,
        const Mat& distCoeffs, const Point3f& zero,
        const Point3f& pb1, const Point3f& pb2, float sqWidth,
        float sqHeight, const vector<Point3f>& whole,
        vector<Point2f>& corners) const;
    void generateBasis(Point3f& pb1, Point3f& pb2) const;
    Point3f generateChessBoardCenter(const Mat& camMat,
        const Size& imgSize) const;
    Mat rvec, tvec;
}; // 结束类声明
}; // 结束命名空间声明
const Size imgSize(800, 600);
const Size brdSize(8, 7);
const size_t brds_num = 20;
template<class T> ostream& operator<<(ostream& out, const Mat_<T>& mat)
{
    for(int j = 0; j < mat.rows; ++j)
        for(int i = 0; i < mat.cols; ++i)
            out << mat(j, i) << " ";
    return out;
}
int main()
{
    // 校准窗口的背景
    Mat background(imgSize, CV_8UC3);
    randu(background, Scalar::all(32), Scalar::all(255));
    GaussianBlur(background, background, Size(5, 5), 2);
    cout << "Done" << endl;
    cout << "初始棋盘建立...";
    ChessBoardGenerator cbg(brdSize);
    cbg.rendererResolutionMultiplier = 4;
```

```cpp
        cout << "Done" << endl;
        /* 相机参数 */
        Mat_<double> camMat(3, 3);
        camMat << 300., 0., background.cols/2.,
            0, 300., background.rows/2., 0., 0., 1.;
        Mat_<double> distCoeffs(1, 5);
        distCoeffs << 1.2, 0.2, 0., 0., 0.;
        cout << "建立棋盘...";
        vector<Mat> boards(brds_num);
        vector<Point2f> tmp;
        for(size_t i = 0; i < brds_num; ++i)
            cout << (boards[i] = cbg(background,
                camMat, distCoeffs, tmp), i) << " ";
        cout << "完成" << endl;
        vector<Point3f> chessboard3D;
        for(int j = 0; j < cbg.cornersSize().height; ++j)
            for(int i = 0; i < cbg.cornersSize().width; ++i)
                chessboard3D.push_back(Point3i(i, j, 0));
        /* init points */
        vector< vector<Point3f> > objectPoints;
        vector< vector<Point2f> > imagePoints;
        cout << endl << "寻找棋盘角...\n";
        for(size_t i = 0; i < brds_num; ++i)
        {
            namedWindow("棋盘");
            imshow("棋盘", boards[i]);
            waitKey(100);
            bool found = findChessboardCorners(boards[i],
                cbg.cornersSize(), tmp);
            if (found)
            {
                imagePoints.push_back(tmp);
                objectPoints.push_back(chessboard3D);
                cout<< i <<"-found\t";
            }
            else
                cout<< i << "-not-found\t";
            // 绘制寻找棋盘角结果
            drawChessboardCorners(boards[i], cbg.cornersSize(),
                Mat(tmp), found);
            imshow("棋盘", boards[i]);
            waitKey(100);
            cout << i << "\n";
        }
        cout << "Done" << endl;
        Mat camMat_est;
        Mat distCoeffs_est;
        vector<Mat> rvecs, tvecs;
        cout << "校准...";
        double rep_err = calibrateCamera(objectPoints, imagePoints,
            imgSize, camMat_est, distCoeffs_est, rvecs, tvecs);
        cout << "Done" << endl;
        cout << endl << "Average Reprojection error: "
            << rep_err/brds_num/cbg.cornersSize().area() << endl;
        cout << "==================================" << endl;
        cout << "Original camera matrix:\n" << camMat << endl;
        cout << "Original distCoeffs:\n" << distCoeffs << endl;
```

```cpp
        cout << "==================================" << endl;
        cout << "Estimated camera matrix:\n"
            << (Mat_<double>&)camMat_est << endl;
        cout << "Estimated distCoeffs:\n"
            << (Mat_<double>&)distCoeffs_est << endl;
        getchar();
        return 0;
}
ChessBoardGenerator::ChessBoardGenerator(const Size& _patternSize) :
sensorWidth(32),
    sensorHeight(24), squareEdgePointsNum(200), min_cos(sqrt(2.f)*0.5f),
cov(0.5),
    patternSize(_patternSize), rendererResolutionMultiplier(4),
tvec(Mat::zeros(1, 3, CV_32F))
{
    Rodrigues(Mat::eye(3, 3, CV_32F), rvec);
}
void cv::ChessBoardGenerator::generateEdge(const Point3f& p1, const
Point3f& p2,
    vector<Point3f>& out) const
{
    Point3f step = (p2 - p1) * (1.f/squareEdgePointsNum);
    for(size_t n = 0; n < squareEdgePointsNum; ++n)
        out.push_back( p1 + step * (float)n);
}
Size cv::ChessBoardGenerator::cornersSize() const
{
    return Size(patternSize.width-1, patternSize.height-1);
}
struct Mult
{
    float m;
    Mult(int mult) : m((float)mult) {}
    Point2f operator()(const Point2f& p)const { return p * m; }
};
void cv::ChessBoardGenerator::generateBasis(Point3f& pb1, Point3f& pb2)
const
{
    RNG& rng = theRNG();
    Vec3f n;
    for(;;)
    {
        n[0] = rng.uniform(-1.f, 1.f);
        n[1] = rng.uniform(-1.f, 1.f);
        n[2] = rng.uniform(-1.f, 1.f);
        float len = (float)norm(n);
        n[0]/=len;
        n[1]/=len;
        n[2]/=len;
        if (fabs(n[2]) > min_cos)
            break;
    }
    Vec3f n_temp = n; n_temp[0] += 100;
    Vec3f b1 = n.cross(n_temp);
    Vec3f b2 = n.cross(b1);
    float len_b1 = (float)norm(b1);
    float len_b2 = (float)norm(b2);
```

```cpp
        pb1 = Point3f(b1[0]/len_b1, b1[1]/len_b1, b1[2]/len_b1);
        pb2 = Point3f(b2[0]/len_b1, b2[1]/len_b2, b2[2]/len_b2);
}
Mat cv::ChessBoardGenerator::generageChessBoard(const Mat& bg,
    const Mat& camMat, const Mat& distCoeffs,
    const Point3f& zero, const Point3f& pb1, const Point3f& pb2,
    float sqWidth, float sqHeight, const vector<Point3f>& whole,
    vector<Point2f>& corners) const
{
    vector< vector<Point> > squares_black;
    for(int i = 0; i < patternSize.width; ++i)
        for(int j = 0; j < patternSize.height; ++j)
            if ( (i % 2 == 0 && j % 2 == 0) || (i % 2 != 0 && j % 2 != 0) )
            {
                vector<Point3f> pts_square3d;
                vector<Point2f> pts_square2d;
                Point3f p1 = zero + (i + 0) * sqWidth * pb1 + (j + 0) *
                            sqHeight * pb2;
                Point3f p2 = zero + (i + 1) * sqWidth * pb1 + (j + 0) *
                            sqHeight * pb2;
                Point3f p3 = zero + (i + 1) * sqWidth * pb1 + (j + 1) *
                            sqHeight * pb2;
                Point3f p4 = zero + (i + 0) * sqWidth * pb1 + (j + 1) *
                            sqHeight * pb2;
                generateEdge(p1, p2, pts_square3d);
                generateEdge(p2, p3, pts_square3d);
                generateEdge(p3, p4, pts_square3d);
                generateEdge(p4, p1, pts_square3d);
                projectPoints( Mat(pts_square3d), rvec, tvec,
                    camMat, distCoeffs, pts_square2d);
                squares_black.resize(squares_black.size() + 1);
                vector<Point2f> temp;
                approxPolyDP(Mat(pts_square2d), temp, 1.0, true);
                transform(temp.begin(), temp.end(),
                    back_inserter(squares_black.back()),
                    Mult(rendererResolutionMultiplier));
            }
    /* calculate corners */
    vector<Point3f> corners3d;
    for(int j = 0; j < patternSize.height - 1; ++j)
        for(int i = 0; i < patternSize.width - 1; ++i)
            corners3d.push_back(zero + (i + 1) *
            sqWidth * pb1 + (j + 1) * sqHeight * pb2);
    corners.clear();
    projectPoints( Mat(corners3d), rvec, tvec,
        camMat, distCoeffs, corners);
    vector<Point3f> whole3d;
    vector<Point2f> whole2d;
    generateEdge(whole[0], whole[1], whole3d);
    generateEdge(whole[1], whole[2], whole3d);
    generateEdge(whole[2], whole[3], whole3d);
    generateEdge(whole[3], whole[0], whole3d);
    projectPoints( Mat(whole3d), rvec, tvec,
        camMat, distCoeffs, whole2d);
    vector<Point2f> temp_whole2d;
    approxPolyDP(Mat(whole2d), temp_whole2d, 1.0, true);
    vector< vector<Point > > whole_contour(1);
```

```cpp
        transform(temp_whole2d.begin(), temp_whole2d.end(),
            back_inserter(whole_contour.front()),
          Mult(rendererResolutionMultiplier));
    Mat result;
    if (rendererResolutionMultiplier == 1)
    {
        result = bg.clone();
        drawContours(result, whole_contour, -1,
        Scalar::all(255), CV_FILLED, CV_AA);
        drawContours(result, squares_black, -1,
        Scalar::all(0), CV_FILLED, CV_AA);
    }
    else
    {
        Mat tmp;
        resize(bg, tmp, bg.size() * rendererResolutionMultiplier);
        drawContours(tmp, whole_contour, -1,
         Scalar::all(255), CV_FILLED, CV_AA);
        drawContours(tmp, squares_black, -1,
         Scalar::all(0), CV_FILLED, CV_AA);
        resize(tmp, result, bg.size(), 0, 0, INTER_AREA);
    }
    return result;
}
Mat cv::ChessBoardGenerator::operator ()(const Mat& bg,
   const Mat& camMat, const Mat& distCoeffs,
   vector<Point2f>& corners) const
{
    cov = min(cov, 0.8);
    double fovx, fovy, focalLen;
    Point2d principalPoint;
    double aspect;
    calibrationMatrixValues( camMat, bg.size(),
      sensorWidth, sensorHeight, fovx, fovy,
      focalLen, principalPoint, aspect);
    RNG& rng = theRNG();
    float d1 = static_cast<float>(rng.uniform(0.1, 10.0));
    float ah = static_cast<float>(rng.uniform(-fovx/2 * cov,
      fovx/2 * cov) * CV_PI / 180);
      float av = static_cast<float>(rng.uniform(-fovy/2 * cov,
      fovy/2 * cov) * CV_PI / 180);
    Point3f p;
    p.z = cos(ah) * d1;
    p.x = sin(ah) * d1;
    p.y = p.z * tan(av);
    Point3f pb1, pb2;
    generateBasis(pb1, pb2);
    float cbHalfWidth = static_cast<float>(norm(p) *
                       sin( min(fovx, fovy) * 0.5 * CV_PI / 180));
    float cbHalfHeight = cbHalfWidth * patternSize.height
       / patternSize.width;
    vector<Point3f> pts3d(4);
    vector<Point2f> pts2d(4);
    for(;;)
    {
        pts3d[0] = p + pb1 * cbHalfWidth + cbHalfHeight * pb2;
        pts3d[1] = p + pb1 * cbHalfWidth - cbHalfHeight * pb2;
```

```cpp
            pts3d[2] = p - pb1 * cbHalfWidth - cbHalfHeight * pb2;
            pts3d[3] = p - pb1 * cbHalfWidth + cbHalfHeight * pb2;
            /* can remake with better perf */
            projectPoints( Mat(pts3d), rvec, tvec,
              camMat, distCoeffs, pts2d);
            bool inrect1 = pts2d[0].x < bg.cols && pts2d[0].y
              < bg.rows && pts2d[0].x > 0 && pts2d[0].y > 0;
            bool inrect2 = pts2d[1].x < bg.cols && pts2d[1].y
              < bg.rows && pts2d[1].x > 0 && pts2d[1].y > 0;
            bool inrect3 = pts2d[2].x < bg.cols && pts2d[2].y
              < bg.rows && pts2d[2].x > 0 && pts2d[2].y > 0;
            bool inrect4 = pts2d[3].x < bg.cols && pts2d[3].y
              < bg.rows && pts2d[3].x > 0 && pts2d[3].y > 0;
            if ( inrect1 && inrect2 && inrect3 && inrect4)
                break;
            cbHalfWidth*=0.8f;
            cbHalfHeight = cbHalfWidth * patternSize.height
              / patternSize.width;
        }
        cbHalfWidth  *= static_cast<float>(patternSize.width)
          / (patternSize.width + 1);
        cbHalfHeight *= static_cast<float>(patternSize.height)
          / (patternSize.height + 1);
        Point3f zero = p - pb1 * cbHalfWidth - cbHalfHeight * pb2;
        float sqWidth  = 2 * cbHalfWidth/patternSize.width;
        float sqHeight = 2 * cbHalfHeight/patternSize.height;
        return generageChessBoard(bg, camMat, distCoeffs, zero,
          pb1, pb2, sqWidth, sqHeight, pts3d, corners);
}
```

执行结果如图 5-2 所示。

图 5-2

5.2 视差

视差（disparity）是关于左右两个图像相对点间的距离，例如，图 5-3 中的 X 点与 OL 连线，你会看到交叉点 XL（图像的投影点），右边交叉点是 XR。如果执行左右两图间每个像素匹配并计算距离，结果的图像就包含有像素间距离的不对称值。

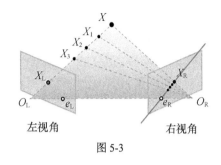

图 5-3

```cpp
#include <stdio.h>
#include <iostream>
#include "opencv2/core/core.hpp"
#include "opencv2/highgui/highgui.hpp"
#include "opencv2/imgproc/imgproc.hpp"
#include "opencv2/calib3d/calib3d.hpp"
using namespace cv;
const char *windowLeft = "Left";
const char *windowRight = "Right";
const char *windowDisparity = "Disparity";
void readme();
int main(int argc, char** argv)
{
    //-- 1. 加载图形文件
    Mat imgLeft = imread("C:\\images\\tsucuba_left.png", 1);
    Mat imgRight = imread("C:\\images\\tsucuba_right.png",1);
    namedWindow(windowLeft, WINDOW_NORMAL);
    imshow(windowLeft, imgLeft);
    namedWindow(windowRight, WINDOW_NORMAL);
    imshow(windowRight, imgRight);
    cvtColor(imgLeft, imgLeft, CV_BGR2GRAY);
    cvtColor(imgRight, imgRight, CV_BGR2GRAY);
    //-- 建立不对称图像
    Mat imgDisparity16S = Mat(imgLeft.rows, imgLeft.cols, CV_16S);
    Mat imgDisparity8U = Mat(imgLeft.rows, imgLeft.cols, CV_8UC1);
    if (!imgLeft.data || !imgRight.data)
    {
        std::cout << " --(!) Error reading images " << std::endl;
        return -1;
    }
    //-- 2. 调用类 StereoBM 的构造函数
    /** 不对称性范围 */
    int ndisparities = 16 * 5;
    /** 区块窗口(block window)大小，必须是奇数 */
    int SADWindowSize = 21;
    StereoBM sbm(StereoBM::BASIC_PRESET,
        ndisparities, SADWindowSize);
    //-- 3. 计算不对称性图像
    sbm(imgLeft, imgRight, imgDisparity16S, CV_16S);
    //-- 检查极端值
    double minVal; double maxVal;
```

```
    minMaxLoc(imgDisparity16S, &minVal, &maxVal);
    printf("Min disp: %f Max value: %f \n", minVal, maxVal);
    //-- 4. 用CV_8UC1值显示成灰度图像
    imgDisparity16S.convertTo(imgDisparity8U, CV_8UC1,
        255 / (maxVal - minVal));
    namedWindow(windowDisparity, WINDOW_NORMAL);
    imshow(windowDisparity, imgDisparity8U);
    waitKey(0);
    return 0;
}
```

程序说明

StereoBM::StereoBM(int preset, int ndisparities=0, int SADWindowSize=21) StereoBM：类构造函数

（1）preset：指定整组算法参数。

- BASIC_PRESET：适合一般相机的参数。
- FISH_EYE_PRESET：适合广角相机的参数。
- NARROW_PRESET：适合窄角相机的参数。

类建构完成，就可更改参数值。

（2）ndisparities：不对称性寻找范围。

（3）SADWindowSize：用来比较区块的线性大小，这个值必须是奇数。

执行结果如图 5-4 所示。

（a）左图

（b）右图
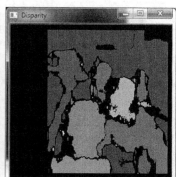
（c）视差图

图 5-4

第 6 章

Feature2d 模块

本模块包括两项功能：第一是图像的特征描述；第二是图像的特征检测。其最终的目的就是进行判断与处理。又因为彩色图色频太多，处理上比较复杂，所以本模块的所有范例都会先转为灰度，这样一来，执行判断与处理就较为单纯。

什么是特征呢？

为什么在图像处理中要找出两个不同的图像之间的匹配点（matching points）？

因为如果我们知道两个图像彼此相关，就可以使用两个图像取得所需的信息：

- 匹配点是指容易判断的特征。
- 特征的特性就是唯一可判断的。

图像特征的类型有以下几类。

- 边线、边缘。
- 角（corners），也称为兴趣点。
- 二进制大型对象（blobs），也称为兴趣区。

为什么角如此特殊？

因为它是两个边的交接处，也代表点在这两个边的方向改变，所以图像的渐变（gradient）有很大的变化，就可用来进行检测。

6.1 特征描述

特征描述的代码如下：

```
#include <stdio.h>
#include <iostream>
#include "opencv2/core/core.hpp"
#include "opencv2/features2d/features2d.hpp"
#include "opencv2/highgui/highgui.hpp"
#include "opencv2/nonfree/features2d.hpp"
using namespace cv;
int main(int argc, char** argv)
{
  Mat img_1 = imread("C:\\images\\tsucuba_left.png",
     CV_LOAD_IMAGE_GRAYSCALE);
  Mat img_2 = imread("C:\\images\\tsucuba_right.png",
     CV_LOAD_IMAGE_GRAYSCALE);
  if (!img_1.data || !img_2.data)
     return -1;
  int minHessian = 400;
  //-- Step 1: 用 SURF 检测器(detector)检测关键点 (keypoints)
  SurfFeatureDetector detector(minHessian);
  std::vector<KeyPoint> keypoints_1, keypoints_2;
  detector.detect(img_1, keypoints_1);
  detector.detect(img_2, keypoints_2);
  //-- Step 2: 用特征向量(feature vectors)计算描述器(descriptors)
  SurfDescriptorExtractor extractor;
  Mat descriptors_1, descriptors_2;
```

```
    extractor.compute(img_1, keypoints_1, descriptors_1);
    extractor.compute(img_2, keypoints_2, descriptors_2);
    //-- Step 3: 用暴力匹配法匹配描述器向量(descriptor vectors)
    BFMatcher matcher(NORM_L2);
    std::vector< DMatch > matches;
    matcher.match(descriptors_1, descriptors_2, matches);
    //-- 绘制匹配相同的
    Mat img_matches;
    drawMatches(img_1, keypoints_1, img_2,
       keypoints_2, matches, img_matches);
    //-- 显示检测相同的
    imshow("Matches", img_matches);
    waitKey(0);
    return 0;
}
```

程序说明

因为在 OpenCV 官网说明文档上，SurfFeatureDetector 类已改成各种类，所以使用下列类函数的说明。OpenCV 一直不断改版，而说明文档会更新成最新的内容。因为 OpenCV 都是能够向上兼容，所以在旧版程序还是可以执行的，这就像 OpenCV 1 仍可执行一样。

1. FeatureDetector::detect(const Mat& image, vector<KeyPoint>& keypoints, const Mat& mask=Mat()) const：在图像中检测关键点（keypoints）

或是：

FeatureDetector::detect(const vector<Mat>& images, vector<vector<KeyPoint>>& keypoints, const vector<Mat>& masks=vector<Mat>()) const

（1）image：输入图像。
（2）keypoints：检测到的关键点。
（3）mask：指明寻找关键点的掩码位置。
（4）images：图像组（set）。

2. drawMatches(const Mat& img1, const vector<KeyPoint>& keypoints1, const Mat& img2, const vector<KeyPoint>& keypoints2, const vector<DMatch>& matches1to2, Mat& outImg, const Scalar& matchColor=Scalar::all(-1), const Scalar& singlePointColor=Scalar::all(-1), const vector<char>& matchesMask=vector<char>(), int flags=DrawMatchesFlags::DEFAULT)：绘制两图之间发现的匹配关键点（keypoints）

或是：

drawMatches(const Mat& img1, const vector<KeyPoint>& keypoints1, const Mat& img2, const vector<KeyPoint>& keypoints2, const vector<vector<DMatch>>& matches1to2, Mat& outImg, const Scalar& matchColor=Scalar::all(-1), const Scalar& singlePointColor=Scalar::all(-1), const vector<vector<char>>& matchesMask=vector<vector<char> >(), int flags=DrawMatchesFlags::DEFAULT)

（1）img1：第一个输入图像。

（2）keypoints1：第一个输入图像的关键点。
（3）img2：第二个输入图像。
（4）keypoints2：第二个输入图像的关键点。
（5）matches1to2：匹配的结果。
（6）outImg：输出图像。
（7）matchColor：匹配两点连接线的颜色。
（8）singlePointColor：找不到匹配点的颜色。
（9）matchesMask：掩码；决定匹配是否要绘制线条，如果值为空，则所有匹配都绘制。
（10）flags：设置绘制特征的旗标。

原图与匹配结果如图 6-1 所示。

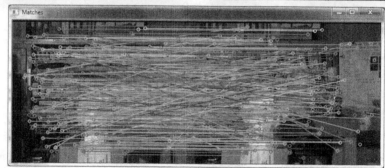

图 6-1

6.2　哈瑞斯角点检测

哈瑞斯角点检测的代码如下：

```
#include "opencv2/highgui/highgui.hpp"
#include "opencv2/imgproc/imgproc.hpp"
#include <iostream>
#include <stdio.h>
#include <stdlib.h>
```

```cpp
using namespace cv;
using namespace std;
/// 声明全局变量
Mat src, src_gray;
int thresh = 200;
int max_thresh = 255;
const char* source_window = "Source image";
const char* corners_window = "Corners detected";
/// 声明函数
void cornerHarris_demo(int, void*);
int main(int, char** argv)
{
    /// 加载图形文件
    src = imread("C:\\images\\building.jpg", 1);
    /// 转成灰度
    cvtColor(src, src_gray, COLOR_BGR2GRAY);
    /// 建立有滑杆的窗口
    namedWindow(source_window, WINDOW_AUTOSIZE);
    createTrackbar("Threshold: ", source_window,
        &thresh, max_thresh, cornerHarris_demo);
    imshow(source_window, src);
    cornerHarris_demo(0, 0);
    waitKey(0);
    return(0);
}
void cornerHarris_demo(int, void*)
{
    Mat dst, dst_norm, dst_norm_scaled;
    dst = Mat::zeros(src.size(), CV_32FC1);
    /// 检测参数
    int blockSize = 2;
    int apertureSize = 3;
    double k = 0.04;
    /// 检测角
    cornerHarris(src_gray, dst, blockSize, apertureSize, k, BORDER_DEFAULT);
    /// Normalizing
    normalize(dst, dst_norm, 0, 255, NORM_MINMAX, CV_32FC1, Mat());
    convertScaleAbs(dst_norm, dst_norm_scaled);
    /// 在角绘制圆圈
    for (int j = 0; j < dst_norm.rows; j++)
    {
        for (int i = 0; i < dst_norm.cols; i++)
        {
            if ((int)dst_norm.at<float>(j, i) > thresh)
            {
                circle(dst_norm_scaled, Point(i, j), 5, Scalar(0), 2, 8, 0);
            }
        }
    }
    /// 显示结果
    namedWindow(corners_window, WINDOW_AUTOSIZE);
    imshow(corners_window, dst_norm_scaled);
}
```

程序说明

cornerHarris(InputArray src, OutputArray dst, int blockSize, int ksize, double k, int borderType=BORDER_DEFAULT)：哈瑞斯（Harris）角点检测

（1）src：输入单色频图像。
（2）dst：哈瑞斯检测结果图像。
（3）blockSize：邻近大小（neighborhood size）。
（4）ksize：Sobel 运算的光圈大小。
（5）k：哈瑞斯检测自由参数（free parameter）。
（6）borderType：像素外推法边缘的类型。

执行结果如图 6-2 所示。

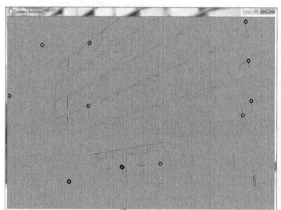

图 6-2

6.3 使用 FLANN 进行特征匹配

使用 FLANN 进行特征匹配的代码如下：

```
#include <stdio.h>
#include <iostream>
#include "opencv2/core/core.hpp"
#include "opencv2/features2d/features2d.hpp"
#include "opencv2/highgui/highgui.hpp"
#include "opencv2/nonfree/features2d.hpp"
using namespace cv;
int main(int argc, char** argv)
{
    Mat img_1 = imread("C:\\images\\lena.jpg", CV_LOAD_IMAGE_GRAYSCALE);
    Mat img_2 = imread("C:\\images\\mixed.jpg", CV_LOAD_IMAGE_GRAYSCALE);
    if (!img_1.data || !img_2.data)
```

```cpp
    {
        std::cout << " --(!) 读图错误 " << std::endl;
        return -1;
    }
    //-- Step 1: 用 SURF 检测器检测关键点
    int minHessian = 400;
    SurfFeatureDetector detector(minHessian);
    std::vector<KeyPoint> keypoints_1, keypoints_2;
    detector.detect(img_1, keypoints_1);
    detector.detect(img_2, keypoints_2);
    //-- Step 2: 用特征向量计算描述器
    SurfDescriptorExtractor extractor;
    Mat descriptors_1, descriptors_2;
    extractor.compute(img_1, keypoints_1, descriptors_1);
    extractor.compute(img_2, keypoints_2, descriptors_2);
    //-- Step 3: 用 FLANN 匹配器匹配描述向量
    FlannBasedMatcher matcher;
    std::vector< DMatch > matches;
    matcher.match(descriptors_1, descriptors_2, matches);
    double max_dist = 0; double min_dist = 100;
    //-- 计算两个关键点间最大与最小距离
    for (int i = 0; i < descriptors_1.rows; i++)
    {
        double dist = matches[i].distance;
        if (dist < min_dist) min_dist = dist;
        if (dist > max_dist) max_dist = dist;
    }
    printf("-- Max dist : %f \n", max_dist);
    printf("-- Min dist : %f \n", min_dist);
    //-- 仅绘制 "好的" 匹配结果
    //-- (距离小于 2 * min_dist 或小的随意 ( 0.02 )
    std::vector< DMatch > good_matches;
    for (int i = 0; i < descriptors_1.rows; i++)
    {
        if (matches[i].distance <= max(2 * min_dist, 0.02))
        {
            good_matches.push_back(matches[i]);
        }
    }
    //-- 仅绘制 "好的" 匹配结果
    Mat img_matches;
    drawMatches(img_1, keypoints_1, img_2, keypoints_2,
        good_matches, img_matches, Scalar::all(-1), Scalar::all(-1),
        vector<char>(), DrawMatchesFlags::NOT_DRAW_SINGLE_POINTS);
    //-- 显示检测匹配结果
    imshow("Good Matches", img_matches);
    for (int i = 0; i < (int)good_matches.size(); i++)
    {
        printf("-- Good Match [%d] Keypoint 1: %d  -- Keypoint 2: %d  \n",
            i, good_matches[i].queryIdx, good_matches[i].trainIdx);
    }
    waitKey(0);
    return 0;
}
```

执行结果如图 6-3 所示。

图 6-3

6.4 使用 Features2D 和 Homography 识别对象

使用 Features2D 和 Homography 识别对象的代码如下：

```
#include <stdio.h>
#include <iostream>
#include "opencv2/core/core.hpp"
#include "opencv2/features2d/features2d.hpp"
#include "opencv2/highgui/highgui.hpp"
#include "opencv2/calib3d/calib3d.hpp"
#include "opencv2/nonfree/features2d.hpp"
using namespace std;
using namespace cv;
int main(int argc, char** argv)
{
  Mat img_object = imread("C:\\images\\lena.jpg", CV_LOAD_IMAGE_GRAYSCALE);
  Mat img_scene = imread("C:\\images\\mixed.jpg", CV_LOAD_IMAGE_GRAYSCALE);
  if (!img_object.data || !img_scene.data)
  {
    std::cout << " --(!) Error reading images " << std::endl;
    return -1;
  }
  //-- Step 1: 用 SURF 检测器检测关键点
  int minHessian = 400;
  SurfFeatureDetector detector(minHessian);
  std::vector<KeyPoint> keypoints_object, keypoints_scene;
  detector.detect(img_object, keypoints_object);
  detector.detect(img_scene, keypoints_scene);
  //-- Step 2: 用特征向量计算描述器
  SurfDescriptorExtractor extractor;
  Mat descriptors_object, descriptors_scene;
  extractor.compute(img_object, keypoints_object, descriptors_object);
  extractor.compute(img_scene, keypoints_scene, descriptors_scene);
  //-- Step 3: 用 FLANN 匹配器匹配描述向量
  FlannBasedMatcher matcher;
  std::vector< DMatch > matches;
  matcher.match(descriptors_object, descriptors_scene, matches);
  double max_dist = 0; double min_dist = 100;
  //-- 计算两点之间最大与最小距离
  for (int i = 0; i < descriptors_object.rows; i++)
```

```cpp
{
    double dist = matches[i].distance;
    if (dist < min_dist) min_dist = dist;
    if (dist > max_dist) max_dist = dist;
}
printf("-- Max dist : %f \n", max_dist);
printf("-- Min dist : %f \n", min_dist);
//-- 仅绘制 "好的" 匹配 (距离小于 3*min_dist)
std::vector< DMatch > good_matches;
for (int i = 0; i < descriptors_object.rows; i++)
{
    if (matches[i].distance < 3 * min_dist)
    {
      good_matches.push_back(matches[i]);
    }
}
Mat img_matches;
drawMatches(img_object, keypoints_object,
    img_scene, keypoints_scene, good_matches,
    img_matches, Scalar::all(-1), Scalar::all(-1),
    vector<char>(), DrawMatchesFlags::NOT_DRAW_SINGLE_POINTS);
//--将对象本地化(Localize)
std::vector<Point2f> obj;
std::vector<Point2f> scene;
for (size_t i = 0; i < good_matches.size(); i++)
{
    //-- 由"好的"匹配取得关键点
    obj.push_back(keypoints_object[good_matches[i].queryIdx].pt);
    scene.push_back(keypoints_scene[good_matches[i].trainIdx].pt);
}
Mat H = findHomography(obj, scene, CV_RANSAC);
//-- 取得 image_1 的角 ( 用来检测的对象 )
std::vector<Point2f> obj_corners(4);
obj_corners[0] = Point(0, 0);
obj_corners[1] = Point(img_object.cols, 0);
obj_corners[2] = Point(img_object.cols, img_object.rows);
obj_corners[3] = Point(0, img_object.rows);
std::vector<Point2f> scene_corners(4);
perspectiveTransform(obj_corners, scene_corners, H);
//-- 在角间画线 (在图 image_2 中匹配到的)
Point2f offset((float)img_object.cols, 0);
line(img_matches, scene_corners[0] + offset,
    scene_corners[1] + offset, Scalar(0, 255, 0), 4);
line(img_matches, scene_corners[1] + offset,
    scene_corners[2] + offset, Scalar(0, 255, 0), 4);
line(img_matches, scene_corners[2] + offset,
    scene_corners[3] + offset, Scalar(0, 255, 0), 4);
line(img_matches, scene_corners[3] + offset,
    scene_corners[0] + offset, Scalar(0, 255, 0), 4);
//-- 显示检测匹配结果
imshow("Good Matches & Object detection", img_matches);
waitKey(0);
return 0;
}
```

程序说明

perspectiveTransform(InputArray src, OutputArray dst, InputArray m) 执行透视矩阵向量的转换

（1）src：输入图像。

（2）dst：输出阵列，与输入图像同大小与类型。

（3）m：3×3 或 4×4 浮点转换矩阵。

执行结果如图 6-4 所示。

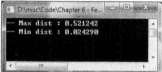

图 6-4

6.5 Shi-Tomasi 角点检测

Shi-Tomasi 角点检测的代码如下：

```
#include "opencv2/highgui/highgui.hpp"
#include "opencv2/imgproc/imgproc.hpp"
#include <iostream>
#include <stdio.h>
#include <stdlib.h>
using namespace cv;
using namespace std;
/// 声明全局变量
Mat src, src_gray;
int maxCorners = 23;
int maxTrackbar = 100;
RNG rng(12345);
const char* source_window = "Image";
/// 声明函数
void goodFeaturesToTrack_Demo(int, void*);
int main(int, char** argv)
{
    /// 加载图形文件并转换成灰度
    src = imread("C:\\images\\board.jpg", 1);
    cvtColor(src, src_gray, COLOR_BGR2GRAY);
```

```cpp
    /// 新建窗口
    namedWindow(source_window, WINDOW_AUTOSIZE);
    /// 新建滑杆来设置角数
    createTrackbar("Max  corners:", source_window,
        &maxCorners, maxTrackbar, goodFeaturesToTrack_Demo);
    imshow(source_window, src);
    goodFeaturesToTrack_Demo(0, 0);
    waitKey(0);
    return(0);
}
void goodFeaturesToTrack_Demo(int, void*)
{
    if (maxCorners < 1) { maxCorners = 1; }
    /// Shi-Tomasi 算法的参数
    vector<Point2f> corners;
    double qualityLevel = 0.01;
    double minDistance = 10;
    int blockSize = 3;
    bool useHarrisDetector = false;
    double k = 0.04;
    /// 复制原图
    Mat copy;
    copy = src.clone();
    /// 角点检测
    goodFeaturesToTrack(src_gray,
        corners,
        maxCorners,
        qualityLevel,
        minDistance,
        Mat(),
        blockSize,
        useHarrisDetector,
        k);
    /// 导出检测到的角 Draw corners detected
    cout << "** 检测到: " << corners.size() << " 个角" << endl;
    int r = 4;
    for (size_t i = 0; i < corners.size(); i++)
    {
        circle(copy, corners[i], r, Scalar(rng.uniform(0, 255),
            rng.uniform(0, 255), rng.uniform(0, 255)), -1, 8, 0);
    }
    /// 显示结果
    namedWindow(source_window, WINDOW_AUTOSIZE);
    imshow(source_window, copy);
}
```

程序说明

goodFeaturesToTrack(InputArray image, OutputArray corners, int maxCorners, double qualityLevel, double minDistance, InputArray mask=noArray(), int blockSize=3, bool useHarrisDetector=false, double k=0.04)：决定图像的强角（strong corners）

（1）image：输入图像。

（2）corners：检测角的输出向量。

（3）maxCorners：返回最大角数。
（4）qualityLevel：角图最低可接受的品质。
（5）minDistance：返回角间可能的最小距离。
（6）mask：可有可无的有兴趣的距离。
（7）blockSize：平均区块大，用来计算每个像素附近衍生共变矩阵（derivation covariation matrix）。
（8）useHarrisDetector：是否使用哈瑞斯检测。
（9）k：哈瑞斯检测的自由参数。

执行结果如图 6-5 所示。

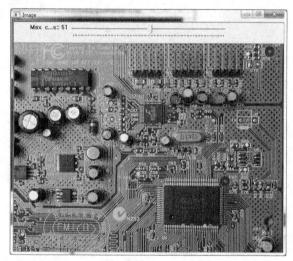

图 6-5

6.6 建立自定义的角点检测

建立自定义的角点检测的代码如下：

```
#include "opencv2/highgui/highgui.hpp"
#include "opencv2/imgproc/imgproc.hpp"
#include <iostream>
#include <stdio.h>
#include <stdlib.h>
using namespace cv;
using namespace std;
/// 声明全局变量
Mat src, src_gray;
Mat myHarris_dst; Mat myHarris_copy; Mat Mc;
Mat myShiTomasi_dst; Mat myShiTomasi_copy;
int myShiTomasi_qualityLevel = 50;
```

```cpp
    int myHarris_qualityLevel = 50;
    int max_qualityLevel = 100;
    double myHarris_minVal; double myHarris_maxVal;
    double myShiTomasi_minVal; double myShiTomasi_maxVal;
    RNG rng(12345);
    const char* myHarris_window = "My Harris corner detector";
    const char* myShiTomasi_window = "My Shi Tomasi corner detector";
    /// 声明函数
    void myShiTomasi_function(int, void*);
    void myHarris_function(int, void*);
    int main(int, char** argv)
    {
        /// 加载图形文件并转换为灰度
        src = imread("C:\\images\\building.jpg", 1);
        cvtColor(src, src_gray, COLOR_BGR2GRAY);
        /// 设置某些参数
        int blockSize = 3; int apertureSize = 3;
        /// Harris 图块
        myHarris_dst = Mat::zeros(src_gray.size(), CV_32FC(6));
        Mc = Mat::zeros(src_gray.size(), CV_32FC1);
        cornerEigenValsAndVecs(src_gray, myHarris_dst,
            blockSize, apertureSize, BORDER_DEFAULT);
        /* 计算 Mc */
        for (int j = 0; j < src_gray.rows; j++)
        {
            for (int i = 0; i < src_gray.cols; i++)
            {
                float lambda_1 = myHarris_dst.at<Vec6f>(j, i)[0];
                float lambda_2 = myHarris_dst.at<Vec6f>(j, i)[1];
                Mc.at<float>(j, i) = lambda_1*lambda_2 - 0.04f*pow((lambda_1 +
                                    lambda_2), 2);
            }
        }
        minMaxLoc(Mc, &myHarris_minVal, &myHarris_maxVal, 0, 0, Mat());
        /* 新建有滑杆的窗口 */
        namedWindow(myHarris_window, WINDOW_AUTOSIZE);
        createTrackbar(" Quality Level:", myHarris_window,
            &myHarris_qualityLevel, max_qualityLevel, myHarris_function);
        myHarris_function(0, 0);
        /// Shi-Tomasi
        myShiTomasi_dst = Mat::zeros(src_gray.size(), CV_32FC1);
        cornerMinEigenVal(src_gray, myShiTomasi_dst,
            blockSize, apertureSize, BORDER_DEFAULT);
        minMaxLoc(myShiTomasi_dst, &myShiTomasi_minVal,
            &myShiTomasi_maxVal, 0, 0, Mat());
        /* 新建有滑杆的窗口 */
        namedWindow(myShiTomasi_window, WINDOW_AUTOSIZE);
        createTrackbar(" Quality Level:", myShiTomasi_window,
            &myShiTomasi_qualityLevel, max_qualityLevel,
            myShiTomasi_function);
        myShiTomasi_function(0, 0);
        waitKey(0);
        return(0);
    }
    void myShiTomasi_function(int, void*)
```

```cpp
{
    myShiTomasi_copy = src.clone();
    if (myShiTomasi_qualityLevel < 1) { myShiTomasi_qualityLevel = 1; }
    for (int j = 0; j < src_gray.rows; j++)
    {
       for (int i = 0; i < src_gray.cols; i++)
       {
          if (myShiTomasi_dst.at<float>(j, i) >
             myShiTomasi_minVal +
             (myShiTomasi_maxVal - myShiTomasi_minVal)
             * myShiTomasi_qualityLevel / max_qualityLevel)
          {
             circle(myShiTomasi_copy, Point(i, j), 4,
               Scalar(rng.uniform(0, 255), rng.uniform(0, 255),
               rng.uniform(0, 255)), -1, 8, 0);
          }
       }
    }
    imshow(myShiTomasi_window, myShiTomasi_copy);
}
void myHarris_function(int, void*)
{
    myHarris_copy = src.clone();
    if (myHarris_qualityLevel < 1) { myHarris_qualityLevel = 1; }
    for (int j = 0; j < src_gray.rows; j++)
    {
       for (int i = 0; i < src_gray.cols; i++)
       {
         if (Mc.at<float>(j, i) >
            myHarris_minVal +
            (myHarris_maxVal - myHarris_minVal) *
             myHarris_qualityLevel / max_qualityLevel)
          {
             circle(myHarris_copy, Point(i, j), 4,
              Scalar(rng.uniform(0, 255), rng.uniform(0, 255),
              rng.uniform(0, 255)), -1, 8, 0);
          }
       }
    }
    imshow(myHarris_window, myHarris_copy);
}
```

程序说明

cornerEigenValsAndVecs(InputArray src, OutputArray dst, int blockSize, int ksize, int borderType=BORDER_DEFAULT)：对图的区块（block）计算 eigen 的值与向量，用来检测角

（1）src：输入单色频图像。

（2）dst：结果图像。

（3）blockSize：平均区块大，用来计算每个像素附近推导协方差矩阵（derivation covariation matrix）。

(4) ksize：用于 Sobel 运算的光圈参数。

(5) borderType：像素外推法边缘的类型。

执行结果如图 6-6 所示。

图 6-6

6.7　在次像素检测角位置

在次像素检测角位置的代码如下：

```
#include "opencv2/highgui/highgui.hpp"
#include "opencv2/imgproc/imgproc.hpp"
#include <iostream>
#include <stdio.h>
#include <stdlib.h>
using namespace cv;
using namespace std;
/// 全域变数
Mat src, src_gray;
int maxCorners = 10;
int maxTrackbar = 25;
RNG rng(12345);
const char* source_window = "Image";
/// 函数
void goodFeaturesToTrack_Demo(int, void*);
int main(int, char** argv)
{
    /// 载入图文件并转换为灰度
    src = imread("C:\\images\\fruits.jpg", 1);
    cvtColor(src, src_gray, COLOR_BGR2GRAY);
    /// 建立窗口
    namedWindow(source_window, WINDOW_AUTOSIZE);
    /// 建立拉杆决定角数
    createTrackbar("Max  corners:", source_window,
        &maxCorners, maxTrackbar, goodFeaturesToTrack_Demo);
    imshow(source_window, src);
```

```cpp
        goodFeaturesToTrack_Demo(0, 0);
        waitKey(0);
        return(0);
    }
    void goodFeaturesToTrack_Demo(int, void*)
    {
        if (maxCorners < 1) { maxCorners = 1; }
        /// Shi-Tomasi 算法的参数
        vector<Point2f> corners;
        double qualityLevel = 0.01;
        double minDistance = 10;
        int blockSize = 3;
        bool useHarrisDetector = false;
        double k = 0.04;
        /// 复制原图
        Mat copy;
        copy = src.clone();
        /// 角点检测
        goodFeaturesToTrack(src_gray,
            corners,
            maxCorners,
            qualityLevel,
            minDistance,
            Mat(),
            blockSize,
            useHarrisDetector,
            k);
        /// 导出检测结果
        cout << "** 检测到: " << corners.size() << " 个角 " << endl;
        int r = 4;
        for (size_t i = 0; i < corners.size(); i++)
        {
            circle(copy, corners[i], r, Scalar(rng.uniform(0, 255),
                rng.uniform(0, 255), rng.uniform(0, 255)), -1, 8, 0);
        }
        /// 显示结果
        namedWindow(source_window, WINDOW_AUTOSIZE);
        imshow(source_window, copy);
        /// 设置需要的参数用来寻找精简的角
        Size winSize = Size(5, 5);
        Size zeroZone = Size(-1, -1);
        /// 反复算法结束准则的类
        TermCriteria criteria =
            TermCriteria(TermCriteria::EPS + TermCriteria::COUNT, 40, 0.001);
        /// 计算精简角的位置
        cornerSubPix(src_gray, corners, winSize, zeroZone, criteria);
        /// 输出
        for (size_t i = 0; i < corners.size(); i++)
        {
            cout << " - 精简角 [" << i << "] ("
                << corners[i].x << ","
                << corners[i].y << ")"
                << endl;
        }
    }
```

程序说明

cornerSubPix(InputArray image, InputOutputArray corners, Size winSize, Size zeroZone, TermCriteria criteria)：精简（refine）角的地点

（1）image：输入图像。
（2）corners：输入角起始坐标与输出的精简坐标（refined coordinates）。
（3）winSize：寻找窗口的一半长度。
（4）zeroZone：搜索区域（search zone）中间死区域（dead region）的一半大小。
（5）criteria：角精简（corner refinement）重复处理结束的准则。

执行结果如图 6-7 所示。

图 6-7

6.8 特征检测

特征检测的代码如下：

```
#include <stdio.h>
#include <iostream>
#include "opencv2/core/core.hpp"
#include "opencv2/features2d/features2d.hpp"
#include "opencv2/nonfree/features2d.hpp"
#include "opencv2/highgui/highgui.hpp"
using namespace cv;
int main(int argc, char** argv)
{
    Mat img_1 = imread("C:\\images\\lena.jpg", CV_LOAD_IMAGE_GRAYSCALE);
    Mat img_2 = imread("C:\\images\\Mixed.jpg", CV_LOAD_IMAGE_GRAYSCALE);
```

```
    if (!img_1.data || !img_2.data)
    {
        std::cout << " --(!) Error reading images " << std::endl;
        return -1;
    }
    //-- Step 1: 用SURF Detector检测关键点
    int minHessian = 400;
    SurfFeatureDetector detector(minHessian);
    std::vector<KeyPoint> keypoints_1, keypoints_2;
    detector.detect(img_1, keypoints_1);
    detector.detect(img_2, keypoints_2);
    //-- 绘制关键点
    Mat img_keypoints_1; Mat img_keypoints_2;
    drawKeypoints(img_1, keypoints_1, img_keypoints_1,
        Scalar::all(-1), DrawMatchesFlags::DEFAULT);
    drawKeypoints(img_2, keypoints_2, img_keypoints_2,
        Scalar::all(-1), DrawMatchesFlags::DEFAULT);
    //-- 显示结果
    imshow("Keypoints 1", img_keypoints_1);
    imshow("Keypoints 2", img_keypoints_2);
    waitKey(0);
    return 0;
}
```

程序说明

drawKeypoints(const Mat& image, const vector<KeyPoint>& keypoints, Mat& outImage, const Scalar& color=Scalar::all(-1), int flags=DrawMatchesFlags::DEFAULT)：绘制关键点

（1）image：输入图像。

（2）keypoints：输入图像的关键点。

（3）outImage：输出图像。

（4）color：关键点的颜色。

（5）flags：绘制特征设置的旗标。

执行结果如图 6-8 所示。

图 6-8

第7章 Video 模块

Video 原本是包含于 HighGUI 模块之中的，但是 Video 的应用比 HighGUI 更为广泛，所以我们用专门的一章来加以介绍。而第 8 章中需要使用到拍摄功能，所以第 7 章特别介绍 OpenCV 如何处理即时图像——拍摄图像。

一般台式计算机或笔记本计算机并没有随机配备摄像头（webcam）设备，读者可以自行选购设备，价格从几十元到几百元不等。设备插入 USB 端口之后，Windows 就会自动安装驱动程序，完成安装后，OpenCV 就可以使用拍摄功能了。

7.1 图像拍摄

图像拍摄的代码如下：

```cpp
#include <opencv\highgui.h>
#include <opencv\cv.h>
#include <iostream>
using namespace cv;
using namespace std;
string intToString(int number){
    std::stringstream ss;
    ss << number;
    return ss.str();
}
int main(int argc, char* argv[])
{
    // 打开编号 0 的摄像头
    VideoCapture cap(0);
    // 打开失败就结束程序
    if (!cap.isOpened())
    {
        cout << "无法启动摄像头" << endl;
        return -1;
    }
    char* windowName = "Webcam Feed";
    // 保存图形文件的文件名
    char filename [50];
    // 文件名序列号起始
    int n = 0;
    // 显示摄像头内容的窗口
    namedWindow(windowName, CV_WINDOW_AUTOSIZE);
    while (1) {
        Mat frame;
        // 从摄像头读取帧 (frame)
        bool bSuccess = cap.read(frame);
        // 检查是否读取成功
        if (!bSuccess)
        {
            cout << "无法从摄像头读取帧" << endl;
            break;
        }
        // 显示读取的帧
        imshow(windowName, frame);
```

```
            // 等待10毫秒是否按键
            switch (waitKey(10))
            {
                // 'Esc' 键被按下('esc' ASCII 的值是 27)
                case 27:
                    // 结束程序
                    return 0;
                    break;
                // 'Enter' 键被按下('Enter' ASCII 的值是 13)
                case 13:
                    sprintf(filename, "C:\\images\\process\\VideoFrame%d.jpg", n++);
                    imwrite(filename, frame);
                    break;
            }
    }
    return 0;
}
```

程序说明

1. VideoCapture cap(0)

这是指计算机连接的摄像头。一般计算机只会有一个摄像头，如果有两个以上，指定第二个的值是 1，依此类推（第一台是 0）。如果将 0 改成计算机硬盘内的图像文件名称，则输入来源就不是计算机连接的摄像头，而是硬盘内的图像文件了。

2. bool VideoCapture::read(Mat& image)：**读取视频帧**

或是：

VideoCapture& VideoCapture::operator>>(Mat& image)

image：视频帧。

这个功能包含 cap.grab()与 cap.retrieve()的结合，也就是调用函数一次会执行两种功能。grab()是从图像文件或摄像头抓取下一帧，retrieve()是将抓到的帧解码再返回解码后的帧。如果 read()没有读到帧就返回 NULL 值，用>>来读取帧就无法取得返回值。

帧的数据类型也是 Mat，所以学会图像处理再将其应用到拍摄也是一样的。读者是否了解为什么照相机拍的照片可以变成视频，就是利用将不间断的帧连接，所以连续不断地拍摄，因而使用无限循环 while (1)来接收每次的视频帧。

本程序是模拟摄像头功能，再通过事件处理让拍摄功能启动。应用的场景更为广泛，例如驾驶超速照相或有闯关拍照等。

7.2 生成视频文件

生成视频文件的代码如下：

```cpp
#include <opencv2/opencv.hpp>
using namespace cv;
using namespace std;
int main()
{
    VideoCapture cap(0);
    if (!cap.isOpened())
    {
        cout << "无法启动摄像头" << endl;
        return -1;
    }
    // 取得图像帧(frames) 大小
    Size S = Size((int)cap.get(CV_CAP_PROP_FRAME_WIDTH),
        (int)cap.get(CV_CAP_PROP_FRAME_HEIGHT));
    // 建立并初始视频存储对象
    VideoWriter put("C:\\images\\process\\output.mpg",
        CV_FOURCC('M', 'P', 'E', 'G'), 30, S);
    if (!put.isOpened())
    {
        cout << "无法产生视频文件" << endl;
        return -1;
    }
    namedWindow("Video");
    // 开始拍摄
    while (char(waitKey(1)) != 'q' && cap.isOpened())
    {
        Mat frame;
        cap >> frame;
        // 检查是否摄像头结束拍摄
        if (frame.empty())
        {
            break;
        }
        imshow("Video", frame);
        put << frame;
    }
    return 0;
}
```

程序说明

本程序会打开摄像头拍摄，直到按下 q 键才结束程序，并将拍摄过程存储成图像文件。因为视频无法通过图书来展示，所以在此省略说明。

1. double VideoCapture::get(int propId)：返回摄像头指定的特性（property）

   propId 的选项有以下几个。
   - CV_CAP_PROP_POS_MSEC：时间戳，该帧在整个视频中的位置。
   - CV_CAP_PROP_POS_FRAMES：帧位置。
   - CV_CAP_PROP_POS_AVI_RATIO：在视频的相对位置；0 为片首，1 为片尾。
   - CV_CAP_PROP_FRAME_WIDTH：帧宽度。
   - CV_CAP_PROP_FRAME_HEIGHT：帧高度。

- CV_CAP_PROP_FPS：帧播放速率。
- CV_CAP_PROP_FOURCC：4 字符编码（4-character code）规则。
- CV_CAP_PROP_FRAME_COUNT：视频帧数。
- CV_CAP_PROP_FORMAT：用 cap.retrieve()取得的帧格式。
- CV_CAP_PROP_MODE：目前抓取的模式。
- CV_CAP_PROP_BRIGHTNESS：亮度，仅用于拍摄。
- CV_CAP_PROP_CONTRAST：对比度，仅用于拍摄。
- CV_CAP_PROP_SATURATION：饱和度，仅用于拍摄。
- CV_CAP_PROP_HUE：色调，仅用于拍摄。
- CV_CAP_PROP_GAIN：增色，仅用于拍摄。
- CV_CAP_PROP_EXPOSURE：曝光，仅用于拍摄。
- CV_CAP_PROP_CONVERT_RGB：布尔标志，表示帧是否该转换成 RGB。
- CV_CAP_PROP_WHITE_BALANCE_U：白平衡设置（Whitebalance setting）的 U 值，只支持 DC1394 V2.x 规格摄像头。
- CV_CAP_PROP_WHITE_BALANCE_V：白平衡设置的 V 值，只支持 DC1394 V2.x 规格摄像头。
- CV_CAP_PROP_RECTIFICATION：用于立体拍摄矫正的旗标。

  只支持 DC1394 V2.x 规格摄像头。
- CV_CAP_PROP_ISO_SPEED 摄像头 ISO 的速度。

  只支持 DC1394 V2.x 规格摄像头。
- CV_CAP_PROP_BUFFERSIZE 内部缓存区存储拍摄图像的容量。

  只支持 DC1394 V2.x 规格摄像头。

2. VideoWrite put(par1, par2, par3, par4)：建立输出图像

（1）par1：图像输出文件名。

（2）par2：图像编码规则；因为规则太多，请读者参考四字符图像编码压缩规格。

（3）par3：每秒的帧数。

（4）par4：图像的大小。

OpenCV 的在线说明文档指出，希望用户用 VideoWriter 的构造函数，所以目前已经查不到 put 的函数声明了，这里将正式的构造函数列于以下：

VideoWriter::VideoWriter(const string& filename, int fourcc, double fps, Size frameSize, bool isColor=true)

（1）filename：图像输出文件名。

（2）fourcc：4 字符图像编码压缩规格。

（3）fps：每秒的帧数。

（4）frameSize：帧的大小。

（5）isColor：如果是非 0 的值，编码默认是彩色的图像，否则以灰度图像处理。

3. cap >> frame

与前一节 cap.read(frame)作用相同，只是另一种语法表现。

4. VideoWriter::write(const Mat& image)：写入视频帧

或是：

VideoWriter& VideoWriter::operator<<(const Mat& image)

image：输出的视频帧

将图像传给存储视频对象。

如果程序中将 VideoCapture cap(0)括号中的数值改成 VideoCapture cap("C:\images\\Megamind.avi")，在视频播放完毕时，while 的循环会自动结束，就如同 4.2 节所示。如此读者应该知道在何种情况下使用 read，而何时使用>>较为恰当。

7.3 指定帧

指定帧的代码如下：

```cpp
#include "opencv2/highgui/highgui.hpp"
#include "opencv2/core/core.hpp"
#include <iostream>
using namespace cv;
using namespace std;
int main()
{
    VideoCapture cap ( "C:\\images\\Megamind.avi" );
    if( ! cap.isOpened () )
        return -1;
    // 帧总数
    int frnb = cap.get(CV_CAP_PROP_FRAME_COUNT);
    cout << "帧总数 = " << frnb << endl;
    Mat frame;
    // 起始帧数
    unsigned int fIdx = 10;
    // 设置帧位置
    cap.set (CV_CAP_PROP_POS_FRAMES , fIdx );
    // 读取帧
    cap >> frame;;
    // 显示帧
    namedWindow("frame", CV_WINDOW_AUTOSIZE);
    imshow("frame ", frame);
    waitKey(3);
    cout << "请输入 0 到 " << frnb - 1 << " 的数字，输入 0 就结束\n" << endl;
    for(;;)
```

```
        {
            cout << "第几个帧？";
            cin >> fIdx;
            if ( fIdx > 0 && fIdx < frnb )
            {
                cap.set ( CV_CAP_PROP_POS_FRAMES , fIdx );
                cap >> frame;;
                imshow("frame ", frame);
                waitKey (3);
            }
            else
                return 0;
        }
        return 0;
    }
```

执行结果如图 7-1 所示。

图 7-1

7.4 移动感知

移动感知的代码如下：

```
#include "opencv2/video/tracking.hpp"
#include "opencv2/imgproc/imgproc.hpp"
#include "opencv2/highgui/highgui.hpp"
#include <iostream>
using namespace cv;
using namespace std;
static void drawOptFlowMap(const Mat& flow, Mat& cflowmap,
    int step, double, const Scalar& color)
{
    for(int y = 0; y < cflowmap.rows; y += step)
        for(int x = 0; x < cflowmap.cols; x += step)
        {
            const Point2f& fxy = flow.at<Point2f>(y, x);
```

```cpp
        // 移动方向线
        line(cflowmap, Point(x,y), Point(cvRound(x+fxy.x),
            cvRound(y+fxy.y)), color);
        // 绿色固定点
        circle(cflowmap, Point(x,y), 2, color, -1);
        }
}
int main(int, char**)
{
    VideoCapture cap(0);
    if( !cap.isOpened() )
        return -1;
    Mat prevgray, gray, flow, cflow, frame;
    namedWindow("flow", 1);
    for(;;)
    {
        cap >> frame;
        cvtColor(frame, gray, COLOR_BGR2GRAY);
        if( prevgray.data )
        {
            // 使用Gunnar Farneback算法计算光流(optical flow)密度
            calcOpticalFlowFarneback(prevgray, gray,
                flow, 0.5, 3, 15, 3, 5, 1.2, 0);
            cvtColor(prevgray, cflow, COLOR_GRAY2BGR);
            // 绘制绿点
            drawOptFlowMap(flow, cflow, 16, 1.5,
                Scalar(0, 255, 0));
            imshow("flow", cflow);
        }
        if(waitKey(30)>=0)
            break;
        // 图像互换
        swap(prevgray, gray);
    }
    return 0;
}
```

程序说明

calcOpticalFlowFarneback(InputArray prev, InputArray next, InputOutputArray flow, double pyr_scale, int levels, int winsize, int iterations, int poly_n, double poly_sigma, int flags)：使用Gunnar Farneback的算法计算光流（optical flow）密度（dense）

（1）prev：输入图像单一色频的前8个位。

（2）next：与prev大小类型相同的图像。

（3）flow：计算的图像流；与prev大小相同但类型是CV_32FC2。

（4）pyr_scale：指定金字塔缩放值，若是0.5表示每层缩减一半。

（5）levels：金字塔层数，若是1表示只有原图。

（6）winsize：平均窗口大小，较大值较易检测移动，但也容易误判。

（7）iterations：金字塔每层重复计算次数。

（8）poly_n：像素周围大小，用来计算每个像素的多项式展开（polynomial expansion）；值越大表示图像越平滑，基本上使用 5 或 7。

（9）poly_sigma：高斯（Gaussian）标准差值，用于多项扩充平滑计算 poly_n = 5，则设为 poly_sigma = 1.1；poly_n = 7，则设为 poly_sigma = 1.5。

（10）flags：旗标值。

- OPTFLOW_USE_INITIAL_FLOW flow：是启始图像流。
- OPTFLOW_FARNEBACK_GAUSSIAN：使用高斯 winsize x winsize 过滤处理。

执行结果如图 7-2 所示。

（a）静止时　　　　　　　　　　　（b）移动时

图 7-2

由图中点变成线，而线条歪斜的方向可以看出移动的方向。

7.5　计算移动时间

计算移动时间的代码如下：

```cpp
#include <iostream>
#include <fstream>
#include "opencv2/video/tracking.hpp"
#include "opencv2/highgui/highgui.hpp"
using namespace cv;
using namespace std;
inline bool isFlowCorrect(Point2f u)
{
    return !cvIsNaN(u.x) && !cvIsNaN(u.y)
        && fabs(u.x) < 1e9 && fabs(u.y) < 1e9;
}
static Vec3b computeColor(float fx, float fy)
```

```cpp
{
    static bool first = true;
    // 颜色转移相对长度:
    // 这些颜色的选定是基于形状相似性(perceptual similarity)
    // 例如红色黄色比黄与绿色好区分色度(shades)
    const int RY = 15;
    const int YG = 6;
    const int GC = 4;
    const int CB = 11;
    const int BM = 13;
    const int MR = 6;
    const int NCOLS = RY + YG + GC + CB + BM + MR;
    static Vec3i colorWheel[NCOLS];
    if (first)
    {
        int k = 0;
        for (int i = 0; i < RY; ++i, ++k)
            colorWheel[k] = Vec3i(255, 255 * i / RY, 0);
        for (int i = 0; i < YG; ++i, ++k)
            colorWheel[k] = Vec3i(255 - 255 * i / YG, 255, 0);
        for (int i = 0; i < GC; ++i, ++k)
            colorWheel[k] = Vec3i(0, 255, 255 * i / GC);
        for (int i = 0; i < CB; ++i, ++k)
            colorWheel[k] = Vec3i(0, 255 - 255 * i / CB, 255);
        for (int i = 0; i < BM; ++i, ++k)
            colorWheel[k] = Vec3i(255 * i / BM, 0, 255);
        for (int i = 0; i < MR; ++i, ++k)
            colorWheel[k] = Vec3i(255, 0, 255 - 255 * i / MR);
        first = false;
    }
    const float rad = sqrt(fx * fx + fy * fy);
    const float a = atan2(-fy, -fx) / (float)CV_PI;
    const float fk = (a + 1.0f) / 2.0f * (NCOLS - 1);
    const int k0 = static_cast<int>(fk);
    const int k1 = (k0 + 1) % NCOLS;
    const float f = fk - k0;
    Vec3b pix;
    for (int b = 0; b < 3; b++)
    {
        const float col0 = colorWheel[k0][b] / 255.f;
        const float col1 = colorWheel[k1][b] / 255.f;
        float col = (1 - f) * col0 + f * col1;
        if (rad <= 1)
            // 以半径增加饱和度
            col = 1 - rad * (1 - col);
        else
            // 超出范围
            col *= .75;
        pix[2 - b] = static_cast<uchar>(255.f * col);
    }
    return pix;
}
static void drawOpticalFlow(const Mat_<Point2f>& flow,
            Mat& dst, float maxmotion = -1)
{
    dst.create(flow.size(), CV_8UC3);
```

```cpp
        dst.setTo(Scalar::all(0));
        // determine motion range:
        float maxrad = maxmotion;
        if (maxmotion <= 0)
        {
            maxrad = 1;
            for (int y = 0; y < flow.rows; ++y)
            {
                for (int x = 0; x < flow.cols; ++x)
                {
                    Point2f u = flow(y, x);
                    if (!isFlowCorrect(u))
                        continue;
                    maxrad = max(maxrad, sqrt(u.x * u.x + u.y * u.y));
                }
            }
        }
        for (int y = 0; y < flow.rows; ++y)
        {
            for (int x = 0; x < flow.cols; ++x)
            {
                Point2f u = flow(y, x);
                if (isFlowCorrect(u))
                    dst.at<Vec3b>(y, x) = computeColor(u.x / maxrad, u.y / maxrad);
            }
        }
}
// 流数据明细的二进制文件格式供参考
// http://vision.middlebury.edu/flow/data/
static void writeOpticalFlowToFile(const Mat_<Point2f>& flow,
        const string& fileName)
{
    static const char FLO_TAG_STRING[] = "PIEH";
    ofstream file(fileName.c_str(), ios_base::binary);
    file << FLO_TAG_STRING;
    file.write((const char*) &flow.cols, sizeof(int));
    file.write((const char*) &flow.rows, sizeof(int));
    for (int i = 0; i < flow.rows; ++i)
    {
        for (int j = 0; j < flow.cols; ++j)
        {
            const Point2f u = flow(i, j);
            file.write((const char*) &u.x, sizeof(float));
            file.write((const char*) &u.y, sizeof(float));
        }
    }
}
int main(int argc, const char* argv[])
{
    VideoCapture cap("C:\\images\\Car.mpg");
    if (!cap.isOpened())
        return -1;
    Mat frame0, frame1;
    // 设置帧位置
    unsigned int fIdx = 55;
```

```
        cap.set(CV_CAP_PROP_POS_FRAMES, fIdx);
        cap >> frame0;
        fIdx = 160;
        cap.set(CV_CAP_PROP_POS_FRAMES, fIdx);
        cap >> frame1;
        imshow("frame0", frame0);
        imshow("frame1", frame1);
        // 图文件从 BGR 转换成灰度
        cvtColor(frame0, frame0, CV_BGR2GRAY);
        cvtColor(frame1, frame1, CV_BGR2GRAY);
        Mat_<Point2f> flow;
        // DUAL TV L1 流算法
        Ptr<DenseOpticalFlow> tvl1 = createOptFlow_DualTVL1();
        const double start = (double)getTickCount();
        tvl1->calc(frame0, frame1, flow);
        const double timeSec = (getTickCount() - start) / getTickFrequency();
        cout << "移动时间 : " << timeSec * 10 << " 秒" << endl;
        Mat out;
        drawOpticalFlow(flow, out);
        // 输出流数据
        if (argc == 2)
            writeOpticalFlowToFile(flow, argv[1]);
        imshow("Flow", out);
        waitKey();
        return 0;
}
```

执行结果如下，视频中的两个帧如图 7-3 所示。

图 7-3

时间计算结果如图 7-4 所示。

色移图如图 7-5 所示。

图 7-4

图 7-5

7.6 即时对象追踪

即时对象追踪的代码如下：

```cpp
#include <opencv\cv.h>
#include <opencv\highgui.h>
using namespace std;
using namespace cv;
// 差异比较敏感值
const static int SENSITIVITY_VALUE = 20;
// 用于模糊平整的大小
const static int BLUR_SIZE = 10;
// 对象位置
int theObject[2] = {0,0};
// 对象矩形框，其中心为对象位置
Rect objectBoundingRectangle = Rect(0,0,0,0);
// 整数转换成字符串
string intToString(int number)
{
    stringstream ss;
    ss << number;
    return ss.str();
}
void searchForMovement(Mat thresholdImage, Mat cameraFeed)
{
    bool objectDetected = false;
    Mat temp;
    thresholdImage.copyTo(temp);
    // 寻找轮廓用的向量值
    vector< vector<Point> > contours;
    vector<Vec4i> hierarchy;
    // 寻找最外围轮廓
    findContours(temp, contours, hierarchy,
        CV_RETR_EXTERNAL,CV_CHAIN_APPROX_SIMPLE );
    // 如果轮廓向量不为空，表示找到对象
    if(contours.size()>0)
        objectDetected=true;
    else
        objectDetected = false;
    if(objectDetected)
    {
        // 最大轮廓是最后的轮廓向量值
        vector< vector<Point> > largestContourVec;
        largestContourVec.push_back(contours.at(contours.size()-1));
        // 最大轮廓的矩形框，并寻找中心点
        objectBoundingRectangle =
            boundingRect(largestContourVec.at(0));
        int xpos = objectBoundingRectangle.x +
                objectBoundingRectangle.width/2;
        int ypos = objectBoundingRectangle.y +
                objectBoundingRectangle.height/2;
        // 对象位置
        theObject[0] = xpos , theObject[1] = ypos;
```

```cpp
    }
    // 对象位置
    int x = theObject[0];
    int y = theObject[1];
    // 垂直线，绿色
    line(cameraFeed, Point(x, y - 10),
        Point(x, y + 10), Scalar(255, 0, 0), 2);
    // 水平线，绿色
    line(cameraFeed, Point(x - 10, y),
        Point(x + 10, y), Scalar(255, 0, 0), 2);
}
int main()
{
    // 按键触发值
    bool objectDetected = false;
    bool debugMode = false;
    bool trackingEnabled = false;
    bool pause = false;
    Mat frame1,frame2;
    Mat grayImage1,grayImage2;
    Mat differenceImage;
    Mat thresholdImage;
    VideoCapture capture;
    while(1)
    {
        // 视频播完再重播
        capture.open("c:/images/54649.mp4");
        if(!capture.isOpened())
        {
            cout<<"读取视频失败\n";
            getchar();
            return -1;
        }
        namedWindow("连续图像差", CV_WINDOW_NORMAL);
        namedWindow("阈值图像", CV_WINDOW_NORMAL);
        namedWindow("图像", CV_WINDOW_NORMAL);
        namedWindow("最终阈值图像", CV_WINDOW_NORMAL);
        // 检查是否最终帧
        while(capture.get(CV_CAP_PROP_POS_FRAMES) <
            capture.get(CV_CAP_PROP_FRAME_COUNT)-1 )
        {
            // 判断是否暂停
            if (!pause)
            {
                // 读取帧
                capture.read(frame1);
                // 转成灰度图像
                cvtColor(frame1, grayImage1, COLOR_BGR2GRAY);
                // 读取帧
                capture.read(frame2);
                // 转成灰度图像
                cvtColor(frame2, grayImage2, COLOR_BGR2GRAY);
            }

            // 比较两连续图像差异并产生强度(intensity)图像
```

```cpp
    absdiff(grayImage1, grayImage2,
        differenceImage);
    // 将强度图像进行阈值处理
    threshold(differenceImage,thresholdImage,
        SENSITIVITY_VALUE,255,THRESH_BINARY);
    // 判断是否为除错模式
    if(debugMode==true)
    {
        // 除错模式，显示结果于窗口
        imshow("连续图像差",differenceImage);
        imshow("阈值图像", thresholdImage);
    }
    else
    {
        // 非除错模式，清除窗口
        destroyWindow("连续图像差");
        destroyWindow("阈值图像");
    }
    // 用模糊处理去除噪声
    blur(thresholdImage, thresholdImage,
        Size(BLUR_SIZE,BLUR_SIZE));
    // 用阈值处理模糊化后的图像，并产生二进制图像
    threshold(thresholdImage, thresholdImage,
        SENSITIVITY_VALUE,255,THRESH_BINARY);
    // 判断是否为除错模式
    if(debugMode==true)
        // 除错模式，显示结果于窗口
        imshow("最终阈值图像",thresholdImage);
    else
        // 非除错模式，清除窗口
        destroyWindow("最终阈值图像");
    // 判断是否为追踪模式
    if(trackingEnabled)
        searchForMovement(thresholdImage,frame1);
    // 显示图像
    imshow("图像",frame1);
    // 检查按键，必须等待10ms
    switch(waitKey(10))
    {
      case 27: //'Esc' 键
        capture.release();
        return 0;
      case 115: //'s' 键
        trackingEnabled = !trackingEnabled;
        break;
      case 100: //'d' 键
        debugMode = !debugMode;
        break;
      case 112: //'p' 键
        pause = !pause;
        break;
    }
}
```

```
        // 视频播放结束
        capture.release();
    }
    return 0;
}
```

执行结果如图 7-6 所示。

(a) 按下 s 键之后　　　　　　(b) 按下 d 键之后

图 7-6

再按下 s 键之后，十字就会消失，如图 7-7 所示。

图 7-7

此图像是使用 Android 手机拍摄的 MPEG4 文件。

ASCII 与键盘对照表如图 7-8 所示。

Dec	Hex	Name	Char	Ctrl-char	Dec	Hex	Char	Dec	Hex	Char	Dec	Hex	Char
0	0	Null	NUL	CTRL-@	32	20	Space	64	40	@	96	60	`
1	1	Start of heading	SOH	CTRL-A	33	21	!	65	41	A	97	61	a
2	2	Start of text	STX	CTRL-B	34	22	"	66	42	B	98	62	b
3	3	End of text	ETX	CTRL-C	35	23	#	67	43	C	99	63	c
4	4	End of xmit	EOT	CTRL-D	36	24	$	68	44	D	100	64	d
5	5	Enquiry	ENQ	CTRL-E	37	25	%	69	45	E	101	65	e
6	6	Acknowledge	ACK	CTRL-F	38	26	&	70	46	F	102	66	f
7	7	Bell	BEL	CTRL-G	39	27	'	71	47	G	103	67	g
8	8	Backspace	BS	CTRL-H	40	28	(	72	48	H	104	68	h
9	9	Horizontal tab	HT	CTRL-I	41	29	)	73	49	I	105	69	i
10	0A	Line feed	LF	CTRL-J	42	2A	*	74	4A	J	106	6A	j
11	0B	Vertical tab	VT	CTRL-K	43	2B	+	75	4B	K	107	6B	k
12	0C	Form feed	FF	CTRL-L	44	2C	,	76	4C	L	108	6C	l
13	0D	Carriage feed	CR	CTRL-M	45	2D	-	77	4D	M	109	6D	m
14	0E	Shift out	SO	CTRL-N	46	2E	.	78	4E	N	110	6E	n
15	0F	Shift in	SI	CTRL-O	47	2F	/	79	4F	O	111	6F	o
16	10	Data line escape	DLE	CTRL-P	48	30	0	80	50	P	112	70	p
17	11	Device control 1	DC1	CTRL-Q	49	31	1	81	51	Q	113	71	q
18	12	Device control 2	DC2	CTRL-R	50	32	2	82	52	R	114	72	r
19	13	Device control 3	DC3	CTRL-S	51	33	3	83	53	S	115	73	s
20	14	Device control 4	DC4	CTRL-T	52	34	4	84	54	T	116	74	t
21	15	Neg acknowledge	NAK	CTRL-U	53	35	5	85	55	U	117	75	u
22	16	Synchronous idle	SYN	CTRL-V	54	36	6	86	56	V	118	76	v
23	17	End of xmit block	ETB	CTRL-W	55	37	7	87	57	W	119	77	w
24	18	Cancel	CAN	CTRL-X	56	38	8	88	58	X	120	78	x
25	19	End of medium	EM	CTRL-Y	57	39	9	89	59	Y	121	79	y
26	1A	Substitute	SUB	CTRL-Z	58	3A	:	90	5A	Z	122	7A	z
27	1B	Escape	ESC	CTRL-[	59	3B	;	91	5B	[	123	7B	{
28	1C	File separator	FS	CTRL-\	60	3C	<	92	5C	\	124	7C	|
29	1D	Group separator	GS	CTRL-]	61	3D	=	93	5D	]	125	7D	}
30	1E	Record separator	RS	CTRL-^	62	3E	>	94	5E	^	126	7E	~
31	1F	Unit separator	US	CTRL-_	63	3F	?	95	5F	_	127	7F	DEL

图 7-8

7.7　播放暂停

播放暂停的代码如下：

```
#include <opencv2/core/core.hpp>
#include <opencv2/highgui/highgui.hpp>
#include <iostream>
using namespace cv;
using namespace std;
int switch_value = 1;
void switch_off_function()
{
};
void switch_on_function()
{
};
static void switch_callback(int, void*)
{
   if (switch_value == 0)
      switch_off_function();
   else
      switch_on_function();
}

int main( int argc, char* argv[] )
{
   VideoCapture cap("C:\\images\\car.mp4");
   if (!cap.isOpened())
```

```
    {
        printf("无法打开 %s\n", "C:\\images\\car.mp4");
        return -1;
    }
    namedWindow( "播放", CV_WINDOW_NORMAL );
    char TrackbarName[10];
    sprintf(TrackbarName, "开关");
    int keep_play = 1;
    createTrackbar(TrackbarName, "播放", &switch_value,
        keep_play, switch_callback);
    Mat frame;
    while( 1 )
    {
        if (switch_value)
        {
            cap >> frame;
            if (frame.empty())
                break;
        }
        imshow("播放", frame);
        if(waitKey(10)==27 ) break;
    }
    return(0);
}
```

本程序展示图像播放暂停功能。

执行结果如图 7-9 所示。

图 7-9

第 8 章

Objdetect 模块

本章是通过级联式的分类器（Cascade Classifier）作为摄影的对象检测。所谓分类器就是利用几百个特定对象的样本，例如脸部或身体相同大小的图像（例如 20×20）来进行训练。当分类器完成训练后，就可以应用到图像中我们感兴趣的区域。分类器输出 1 表示该区域可能有我们需要的对象，例如脸部或车体；输出若是 0 表示没有我们感兴趣的对象。要在整个图像中找到我们想要的对象，就必须将分类器在图像内每个区域移动寻找。所以分类器设计成通过更改大小来寻找我们有兴趣的对象的大小，这比改变图像大小简单又有效率。所以要寻找图像中未知大小的对象，就必须使用不同大小的分类器寻找。

"级联式"在分类器上的意思，是由许多简单的分类器或称阶段（stages）组成的合成分类器（resultant classifier）。以连续的方式应用于我们有兴趣的区域，直到某些阶段的候选（candidate）被拒绝或所有阶段都通过。

"提升式（boosted）"的意思，是分类器在级联式每个阶段的自我复杂化，应用 4 种提升技术中的一种。以加权推举的方式，来建立基本的分类器。目前对 Discrete Adaboost、Real Adaboost、Gentle Adaboost 与 Logitboost 都有支持。

基本分类器是最少两种选择的决策树分类器，Haar-like 特征用来输入给基本分类器并加以计算。目前算法用到下面的 Haar-like 特征。

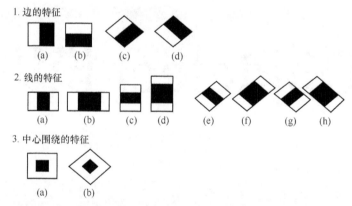

用于特定分类器的特征是由形状（1a、2b 等）、感兴趣区域的位置、比例（scale）来指定的。例如，线特征 2c 情况的响应，是计算黑色条纹乘以 3，然后再除以在涵盖所有特征的矩形区内像素总和与全部像素总和之间的差。

级联式类分类

级联式类分类的代码如下：

```
#include "opencv2/objdetect/objdetect.hpp"
#include "opencv2/highgui/highgui.hpp"
```

```cpp
#include "opencv2/imgproc/imgproc.hpp"
#include <iostream>
#include <stdio.h>
using namespace std;
using namespace cv;
/** 函数声明 */
void detectAndDisplay(Mat frame);
/** 全局变量 */
String face_cascade_name = "C:\\OpenCV\\sources\\data\\haarcascades\\
haarcascade_frontalface_alt.xml";
String eyes_cascade_name = "C:\\OpenCV\\sources\\data\\haarcascades\\
haarcascade_eye_tree_eyeglasses.xml";
// 级联式分类器声明
CascadeClassifier face_cascade;
CascadeClassifier eyes_cascade;
String window_name = "脸部检测";
/** 主程序 */
int main(void)
{
    VideoCapture capture;
    Mat frame;
    //-- 1. 载入全局变量
    if (!face_cascade.load(face_cascade_name))
    {
        printf("--(!) 无法载入脸部 cascade\n");
        return -1;
    };
    if (!eyes_cascade.load(eyes_cascade_name))
    {
        printf("--(!) 无法载入眼睛 cascade\n");
        return -1;
    };
    //-- 2. 打开摄像头并读取摄影图像
    capture.open(0);
    if (!capture.isOpened())
    {
        printf("--(!) 无法打开摄像头\n");
        getchar();
        return -1;
    }
    //     读取帧
    while (capture.read(frame))
    {
        if (frame.empty())
        {
            printf(" --(!) 无法读取帧!");
            getchar();
            break;
        }
        //-- 3. 用级联式类进行摄影帧处理
        detectAndDisplay(frame);
        int c = waitKey(10);
        if ((char)c == 27) { break; } // Esc 键
    }
    return 0;
}
/** 函数 detectAndDisplay 实体 */
```

```cpp
void detectAndDisplay(Mat frame)
{
    vector<Rect> faces;
    Mat frame_gray;
    // 将帧转成灰度
    cvtColor(frame, frame_gray, COLOR_BGR2GRAY);
    // 灰度帧进行直条式的评等化
    equalizeHist(frame_gray, frame_gray);
    //-- 检测脸部
    face_cascade.detectMultiScale(frame_gray,
        faces, 1.1, 2, 0 | CASCADE_SCALE_IMAGE, Size(30, 30));
    for (size_t i = 0; i < faces.size(); i++)
    {
        Point center(faces[i].x + faces[i].width / 2,
            faces[i].y + faces[i].height / 2);
        ellipse(frame, center, Size(faces[i].width / 2,
            faces[i].height / 2), 0, 0, 360,
            Scalar(255, 0, 255), 4, 8, 0);
        printf("绘制脸部, 脸部的大小 %d.............\n", faces.size());
        Mat faceROI = frame_gray(faces[i]);
        vector<Rect> eyes;
        //-- 在脸部检测眼睛
        eyes_cascade.detectMultiScale(faceROI, eyes,
            1.1, 2, 0 | CASCADE_SCALE_IMAGE, Size(30, 30));
        for (size_t j = 0; j < eyes.size(); j++)
        {
            Point eye_center(faces[i].x + eyes[j].x + eyes[j].width / 2,
                faces[i].y + eyes[j].y + eyes[j].height / 2);
            int radius = cvRound((eyes[j].width + eyes[j].height)*0.25);
            circle(frame, eye_center, radius, Scalar(255, 0, 0), 4, 8, 0);
            printf("绘制眼睛, 眼睛数量 %d\n\n\n", eyes.size());
        }
    }
    //-- 显示结果
    imshow(window_name, frame);
}
```

程序说明

1. bool CascadeClassifier::load(const string& filename)：由文档载入级联式类

    filename：载入级联式类的文档名。

2. CascadeClassifier::detectMultiScale(const Mat& image, vector<Rect>& objects, double scaleFactor=1.1, int minNeighbors=3, int flags=0, Size minSize=Size(), Size maxSize=Size())：在输入图像中检测不同大小的对象

    （1）image：输入图像。

    （2）objects：矩形向量，每个矩型包含检测到的对象。

    （3）scaleFactor：在每次图像缩放时指定图像变更的大小。

    （4）minNeighbors：指定每个候选矩形（candidate rectangle）应该拥有多少的邻近者

(neighbors)。

（5）flags：新版已不用此旗标。

（6）minSize：对象可能最小的大小，比此数值小则忽略。

（7）maxSize：对象可能最大的大小，比此数值大则忽略。

执行结果如图 8-1 所示。

图 8-1

本程序是摄像头操作，为了方便显示结果，所以更改以图片呈现结果。读者只要将 capture.open(0)改成 capture.open("C:\\images\\lena.jpg")即可。

在目录 C:/opencv/sources/data 中，就是检测脸部各器官的相关文档。

第 9 章

ML 模块

机器学习（Machine Learning）这个名词现在应用很广泛，一般来说就是通过数据训练与测试达到对数据要处理的结果。在大数据分析上，就是要对数据模式进行判断；在图像处理上，则可能是图形的判读或识别。所以用强有力机器学习（Machine Learning）的算法可以进行统计分类、回归分析与数据聚类（clustering）。

一般数据是不具有标签（label）。例如脸的轮廓本身没有标签，但加入年龄时就具有标签。如果数据具有标签称为数据被监督（supervised），否则称没被监督（unsupervised）。而有标签的数据是可以被分类的（categorical），例如根据年龄或性别分类。如果数据是已分类的就称为数据已分类（classification），所以数据回归分析就是做数据分类。

在 OpenCV 中，ML 模块是一组统计分类、回归分析、数据聚类的类与统计模式函数，被所有 ML 算法所使用。OpenCV 的算法都是有区分的（discriminative）算法，也就是结果会给数据可能的标签，虽然分别不是很清楚，但已经可以用来进行预测。

ML 的算法如表 9-1 所示。

表 9-1

算 法	说 明
Statistical Models	统计模式
Normal Bayes Classifier （正态贝叶斯分类器）	假设每个类特征向量是正态分布，所以整个数据分布假设为高斯（Gaussian）混合
K-Nearest Neighbors （K 近邻法）	取得样品并预测新样品，使用表决、权重计算来分析 K 个邻近值，这个方法也称为由样品学习法
Support Vector Machine (SVM) （支持向量机）	支持向量机原先是用于二元分类优化，现已扩展为回归分析与数据聚类
Decision Trees （决策树）	主要是分类与回归树的演算
Boosting （提升法）	是计算二者之间函数的关系，F:y=F(x)； x 是输入值，y 是输出值
Gradient Boosted Trees (GBT) （梯度提升树）	是提升法更广泛的应用
Random Trees （随机树）	主要是处理分类与回归分析
Extremely Randomized Trees （极端随机树）	此算法与随机树相似，使用相同的训练数据组，并找出树中最佳分叉点
Expectation Maximization （期待最大值）	评估多变概率密度函数（multivariate probability density）的参数

续表

算 法	说 明
Neural Networks （神经网络）	也称为多层认知（Multi-Layer Perception，MLP），分为输入层、输出层与隐藏层，各层间通过类神经连接 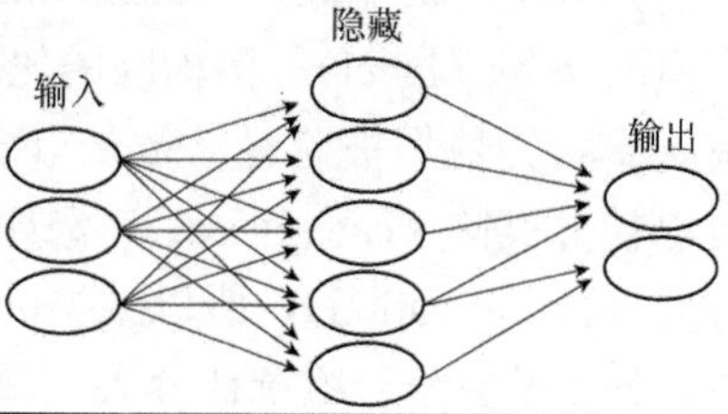
ML Data	是针对 csv 文档格式的数据作预测

另外，还有 Mahalanobis 与 K 均值(K-means)算法是放在 Core 模块。

Mahalanobis	测量向量数据间的距离
K-means	用 K 中心值代表数据分布，K 是演算前确定的
K 均值	主要用来进行数据聚类

OpenCV 关于 ML 常用的代码参见表 9-2。

表 9-2

代 码	说 明

save(
  const char* filename,
  const char* name = 0)
	用 XML 或 YMAL 存储学习的模式

load(
  const char* filename,
  const char* name = 0)
	载入 XML 或 YMAL 的存储学习模式

clear()	清除内存内容 一般载入学习模式前，会先执行此动作
bool train(....)	根据不同算法会有不同的训练函数， 用来取得数据模式
	只要在 OpenCV 的说明文档，于搜索栏输入 bool train(就会列出所有正确的函数内容
float predict(...)	根据不同算法会有不同的预估函数，用来取得数据标签（label），或新训练点的值
	只要在 OpenCV 的说明文档，在搜索栏输入 float predict(就会列出所有正确的函数内容

9.1 支持向量机的介绍

支持向量机（Support Vector Machine，SVM）是一种由分开的超平面（hyperplane）所定义有区分的分类法，也就是提供有标签训练的数据。此算法就会产生最佳超平面。

例如，图 9-1 以直线分开的两组数据，直线代表分类，而直线有许多种画法，则哪一条是最适当的分类？

我们可以靠直觉来增加一些条件，来评估哪些直线比较好，例如不能太靠近点。这样，SVM 算法就能够根据所赋予的条件计算出超平面，也就是图 9-2 中间的实线。实线与虚线之间值的两倍就称为边距（margin），也就是两条虚线间距。

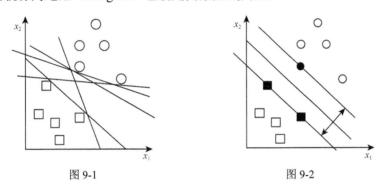

图 9-1   图 9-2

至于超平面是如何进行计算，涉及理论较深，因此不适合初学者，并且算法的函数会自动计算，所以我们在此省略。

```
#include <opencv2/core/core.hpp>
#include <opencv2/highgui/highgui.hpp>
#include <opencv2/ml/ml.hpp>
using namespace cv;
int main()
{
    // 要呈现的数据
    int width = 512, height = 512;
    Mat image = Mat::zeros(height, width, CV_8UC3);
    // 标签
    float labels[4] = { 1.0, -1.0, -1.0, -1.0 };
    Mat labelsMat(4, 1, CV_32FC1, labels);
    // 设置训练数据
    float trainingData[4][2] = { { 501, 10 }, { 255, 10 },
                                 { 501, 255 }, { 10, 501 } };
    Mat trainingDataMat(4, 2, CV_32FC1, trainingData);
    // 设置 SVM 参数
    CvSVMParams params;
    params.svm_type = CvSVM::C_SVC;
    params.kernel_type = CvSVM::LINEAR;
```

```
        params.term_crit = cvTermCriteria(CV_TERMCRIT_ITER, 100, 1e-6);
        // 训练 SVM
    CvSVM SVM;
    SVM.train(trainingDataMat, labelsMat, Mat(), Mat(), params);
    Vec3b green(0, 255, 0), blue(255, 0, 0);
        // 显示 SVM 的决定区
    for (int i = 0; i < image.rows; ++i)
        for (int j = 0; j < image.cols; ++j)
        {
           Mat sampleMat = (Mat_<float>(1, 2) << j, i);
           float response = SVM.predict(sampleMat);
           if (response == 1)
              image.at<Vec3b>(i, j) = green;
           else if (response == -1)
              image.at<Vec3b>(i, j) = blue;
        }
        // 显示训练数据
    int thickness = -1;
    int lineType = 8;
    circle(image, Point(501, 10), 5, Scalar(0, 0, 0),
       thickness, lineType);
    circle(image, Point(255, 10), 5, Scalar(255, 255, 255),
       thickness, lineType);
    circle(image, Point(501, 255), 5, Scalar(255, 255, 255),
       thickness, lineType);
    circle(image, Point(10, 501), 5, Scalar(255, 255, 255),
       thickness, lineType);
        // 显示支持向量
    thickness = 2;
    lineType = 8;
    int c = SVM.get_support_vector_count();
    for (int i = 0; i < c; ++i)
    {
       const float* v = SVM.get_support_vector(i);
       circle(image, Point((int)v[0], (int)v[1]), 6,
         Scalar(128, 128, 128), thickness, lineType);
    }
        // 显示结果
    imshow("SVM Simple Example", image);
    waitKey(0);
}
```

程序说明

1. bool CvSVM::train(const Mat& trainData, const Mat& responses, const Mat& varIdx=Mat(), const Mat& sampleIdx=Mat(), CvSVMParams params=CvSVMParams())：训练 SVM

或是：

bool CvSVM::train(const CvMat* trainData, const CvMat* responses, const CvMat* varIdx=0, const CvMat* sampleIdx=0, CvSVMParams params=CvSVMParams())：所有参数都定义在 CvSVMParams 结构中

2. CvSVMParams::CvSVMParams(int svm_type, int kernel_type, double degree, double gamma, double coef0, double Cvalue, double nu, double p, CvMat* class_weights, CvTermCriteria term_crit)：SVM 训练参数设置

(1) svm_type：SVM 公式类型。
- C_SVCC：支持向量分类（Suport Vector Classification）。
- NU_SVC：支持向量分类。
- ONE_CLASS：分布预估（distribution estimation）。
- EPS_SVR：支持向量回归（Support Vector Regression）。
- NU_SVR：支持向量回归。

(2) kernel_type-SVM：核类型。
- LINEAR：线性核（Linear kernel）。
- POLY：多项式核（Polynomial kernel）。
- RBF：星形基本函数（radial basis function）。
- SIGMOID：S 形核（Sigmoid kernel）。

(3) degree：多项式核函数的度数（degree）。
(4) gamma：(POLY/RBF/SIGMOID)核函数的γ值。
(5) coef0：(POLY/SIGMOID)核函数的系数值。
(6) Cvalue：SVM 优化问题（optimization problem）的 C 值。
(7) nu：SVM 优化问题（optimization problem）的 v 值。
(8) p：SVM 优化问题（optimization problem）的 e 值。
(9) class_weights：C_SVC 问题选择性权重。
(10) term_cri：重复 SVM 训练程序结束的要件。

3. float CvSVM::predict(const Mat& sample, bool returnDFVal=false) const：预估输入样本后的反应

或是：

float CvSVM::predict(const CvMat* sample, bool returnDFVal=false) const

或是：

float CvSVM::predict(const CvMat* samples, CvMat* results) const

(1) sample：输入的预测样本，或是输入的样本用来预测。
(2) returnDFVal：返回值的形态；如果为 true 并且问题是二级分类（2-class classification），将返回精准的函数值，即到边距（margin）的距离；否则返回分类标签（class label）或是预估回归函数值。
(3) results：相对样本输出的预估回应（prediction responses）。

4. int CvSVM::get_support_vector_count() const：取得支持向量的数量

   或是：

const float* CvSVM::get_support_vector(int i) const：取得特定的支持向量

   i：特定支持向量的索引。

   执行结果如图 9-3 所示。

图 9-3

- 用不同颜色的点显示训练数据不同的类。
- 区分蓝绿亮色块的线（即右上角的斜线）就是超平面。
- SVM 用白点在训练数据的周围呈现。

9.2 非线性可分开数据的支持向量机

非线性可分开数据的支持向量机的代码如下：

```
#include <iostream>
#include <opencv2/core/core.hpp>
#include <opencv2/highgui/highgui.hpp>
#include <opencv2/ml/ml.hpp>
// 每个课程训练样本数
#define NTRAINING_SAMPLES 100
#define FRAC_LINEAR_SEP 0.9f
using namespace cv;
using namespace std;
int main()
{
    // 要显示的数据
    const int WIDTH = 512, HEIGHT = 512;
```

```cpp
Mat I = Mat::zeros(HEIGHT, WIDTH, CV_8UC3);
//-------- 1. 随机设置训练数据 ------------
Mat trainData(2 * NTRAINING_SAMPLES, 2, CV_32FC1);
Mat labels(2 * NTRAINING_SAMPLES, 1, CV_32FC1);
// 产生随机数
RNG rng(100);
// 设置线性训练数据
int nLinearSamples =
    (int)(FRAC_LINEAR_SEP * NTRAINING_SAMPLES);
// 产生类 1 的随机点
Mat trainClass = trainData.rowRange(0, nLinearSamples);
// 点在[0, 0.4) 内的 x 坐标
Mat c = trainClass.colRange(0, 1);
rng.fill(c, RNG::UNIFORM, Scalar(1), Scalar(0.4 * WIDTH));
// 点在[0, 1) 内的 y 坐标
c = trainClass.colRange(1, 2);
rng.fill(c, RNG::UNIFORM, Scalar(1), Scalar(HEIGHT));
// 产生类 2 的随机点
trainClass = trainData.rowRange(2 * NTRAINING_SAMPLES - nLinearSamples,
    2 * NTRAINING_SAMPLES);
// 点在[0.6, 1] 内的 x 坐标
c = trainClass.colRange(0, 1);
rng.fill(c, RNG::UNIFORM, Scalar(0.6*WIDTH), Scalar(WIDTH));
// 点在[0, 1) 内的 y 坐标
c = trainClass.colRange(1, 2);
rng.fill(c, RNG::UNIFORM, Scalar(1), Scalar(HEIGHT));
//------------- 设置非线性训练数据 ---------------
//类1 与 2 产生随机点
trainClass = trainData.rowRange(nLinearSamples,
    2 * NTRAINING_SAMPLES - nLinearSamples);
// 点在[0.4, 0.6) 内的 x 坐标
c = trainClass.colRange(0, 1);
rng.fill(c, RNG::UNIFORM, Scalar(0.4*WIDTH), Scalar(0.6*WIDTH));
// 点在[0, 1) 内的 y 坐标
c = trainClass.colRange(1, 2);
rng.fill(c, RNG::UNIFORM, Scalar(1), Scalar(HEIGHT));
//---------- 设置类标签 --------------------
// 类 1
labels.rowRange(0, NTRAINING_SAMPLES).setTo(1);
// 类 2
labels.rowRange(NTRAINING_SAMPLES, 2 * NTRAINING_SAMPLES).setTo(2);
//------------- 2. 设置支持向量参数 ---------
CvSVMParams params;
params.svm_type = SVM::C_SVC;
params.C = 0.1;
params.kernel_type = SVM::LINEAR;
params.term_crit = TermCriteria(CV_TERMCRIT_ITER, (int)1e7, 1e-6);
//-------------- 3. 训练 svm ------------------
cout << "开始训练" << endl;
CvSVM svm;
svm.train(trainData, labels, Mat(), Mat(), params);
cout << "结束训练" << endl;
//---------- 4. 显示决定区 (decision regions) ----------
Vec3b green(0, 100, 0), blue(100, 0, 0);
for (int i = 0; i < I.rows; ++i)
```

```cpp
        for (int j = 0; j < I.cols; ++j)
        {
            Mat sampleMat = (Mat_<float>(1, 2) << i, j);
            float response = svm.predict(sampleMat);
            if (response == 1)
                I.at<Vec3b>(j, i) = green;
            else if (response == 2)
                I.at<Vec3b>(j, i) = blue;
        }
    //-------------- 5. 显示训练数据 ----------------
    int thick = -1;
    int lineType = 8;
    float px, py;
    // 类 1
    for (int i = 0; i < NTRAINING_SAMPLES; ++i)
    {
        px = trainData.at<float>(i, 0);
        py = trainData.at<float>(i, 1);
        circle(I, Point((int)px, (int)py), 3,
            Scalar(0, 255, 0), thick, lineType);
    }
    // 类 2
    for (int i = NTRAINING_SAMPLES; i <2 * NTRAINING_SAMPLES; ++i)
    {
        px = trainData.at<float>(i, 0);
        py = trainData.at<float>(i, 1);
        circle(I, Point((int)px, (int)py), 3,
            Scalar(255, 0, 0), thick, lineType);
    }
    //-------------- 6. 显示支持向量 ------------------
    thick = 2;
    lineType = 8;
    int x = svm.get_support_vector_count();
    for (int i = 0; i < x; ++i)
    {
        const float* v = svm.get_support_vector(i);
        circle(I, Point((int)v[0], (int)v[1]), 6,
            Scalar(128, 128, 128), thick, lineType);
    }
    imshow("SVM 用于非线性训练数据", I);
    waitKey(0);
}
```

灰色圆圈就是 SVM 结果，如图 9-4 所示。

图 9-4

第 10 章

Contrib 模块

探索视网膜效果并用来进行图像处理

探索视网膜效果并用来进行图像处理的代码如下：

```cpp
#include <opencv2/core/core.hpp>
#include <opencv2/highgui/highgui.hpp>
#include <iostream>
#include <cstring>
using namespace cv;
using namespace std;
#include "opencv2/opencv.hpp"
int main(int argc, char* argv[]) {
    // 视网膜 log 取样处理
    bool useLogSampling = true;
    string inputMediaType = "-image";
    // 声明视网膜输入缓存区
    Mat inputFrame;
    // 如果使用摄像头就以此做为视网膜输入缓存区
    VideoCapture videoCapture;
    //////////////////////////////////////////////////////////////////
    // 检查输入类型
    if (!strcmp(inputMediaType.c_str(), "-image"))
    {
        // 加载图形文件
        inputFrame = imread("C:\\images\\scene.jpg", 1);
    }
    else
        if (!strcmp(inputMediaType.c_str(), "-video"))
        {
            videoCapture.open(0);
            videoCapture >> inputFrame;
        }
    if (inputFrame.empty())
        return -1;
    //////////////////////////////////////////////////////////////////
    // 因为视网膜处理可能会有错误，所以使用 try/catch 比较安全，不会死机
    try
    {
        // 用标准参数设置来建立视网膜实例(instance)
        Ptr<Retina> myRetina;
        // 如果最后参数使用 log，则启动 log 取样
        if (useLogSampling)
        {
            myRetina = new Retina(inputFrame.size(),
                true, RETINA_COLOR_BAYER, true, 2.0, 10.0);
        }
        else
            // 传统视网膜处理
            myRetina = new Retina(inputFrame.size());
        // 存储默认视网膜参数
```

```cpp
    myRetina->write("C:\\images\\process\\RetinaDefaultParameters.xml");
    // 加载参数，不加载也会使用默认值
    myRetina->setup("C:\\images\\process\\RetinaDefaultParameters.xml");
    // 重设视网膜缓存区，如同闭眼一段长时间
    myRetina->clearBuffers();
    // 声明视网膜输出缓存区
    Mat retinaOutput_parvo;
    Mat retinaOutput_magno;
    while (1)
    {
      // 确认是否具有摄像头并读取摄影图像，否则使用原图
      if (videoCapture.isOpened())
        videoCapture >> inputFrame;
      // 对摄影图像进行视网膜过滤
      myRetina->run(inputFrame);
      // 采集视网膜过滤结果
      // 视网膜详尽颜色通道(details channel)
      myRetina->getParvo(retinaOutput_parvo);
      // 视网膜移动颜色通道(motion channel)
      myRetina->getMagno(retinaOutput_magno);
      // 显示视网膜过滤结果
      imshow("input", inputFrame);
      imshow("Parvo", retinaOutput_parvo);
      imshow("Magno", retinaOutput_magno);
      if (char(waitKey(1)) == 'q') break;
    }
  }
  catch (Exception e)
  {
    cerr << "使用 Retina 错误: " << e.what() << std::endl;
  }
  return 0;
}
```

程序说明

1. Retina::Retina(Size inputSize)：视网膜构造函数

   或是：

Retina::Retina(Size inputSize, const bool colorMode, RETINA_COLORSAMPLINGMETHOD colorSamplingMethod=RETINA_COLOR_BAYER, const bool useRetinaLogSampling=false, const double reductionFactor=1.0, const double samplingStrenght=10.0)

（1）inputSize：输入摄影图像大小。

（2）colorMode：选择的处理模式，有或没有颜色。

（3）colorSamplingMethod：指定颜色取样的种类。

- RETINA_COLOR_RANDOM：每个像素是 R、G 或 B 的随机选取。
- RETINA_COLOR_DIAGONAL：以 RGBRGBRGB 进行颜色取样。

第二行是 BRGBRGBRG。

第三行是 GBRGBRGBR。
- RETINA_COLOR_BAYER：标准拜耳（bayer）取样。

（4）useRetinaLogSampling：启用视网膜取样记录；此值为 true 时，接下来的两个参数才会使用。

（5）reductionFactor：指定高分辨率输出摄影图像降低因子（reduction factor），但没有精确度流失。

（6）samplingStrenght：指定采用日志级别（log scale）的强度。

2. Retina::write(std::string fs) const：将视网膜参数写入 xml 或 yml 格式的文件

或是：

Retina::write(FileStorage& fs) const

Fs：xml 或 yml 文档名。

3. Retina::setup(std::string retinaParameterFile="", const bool applyDefaultSetupOnFailure=true)：调整视网膜参数设置

或是：

Retina::setup(FileStorage& fs, const bool applyDefaultSetupOnFailure=true)

或是：

Retina::setup(RetinaParameters newParameters)

（1）retinaParameterFile：参数文件名。

（2）applyDefaultSetupOnFailure：如果值为 true，发生错误时将会丢出错误。

（3）fs：打开函具有参数的 Filestorage。

（4）newParameters：以新目标配置（target configuration）更新的参数结构（parameters structures）。

4. Retina::clearBuffers()：清除视网膜缓冲区（就像眼睛长时间闭着再张开）

5. Retina::run(const Mat& inputImage)：将视网膜功能应用到图像

inputImage：输入图像。

6. Retina::getParvo(Mat& retinaOutput_parvo)：视网膜详尽颜色通道的访问方法(accessor)

或是：

Retina::getParvo(std::valarray<float>& retinaOutput_parvo)

或是：

const std::valarray<float>& Retina::getParvo() const

retinaOutput_parvo：输出缓冲区。

7. Retina::getMagno(Mat& retinaOutput_magno)：视网膜移动颜色通道的访问方法

   或是：

Retina::getMagno(std::valarray<float>& retinaOutput_magno)

   或是：

const std::valarray<float>& Retina::getMagno() const

retinaOutput_magno：输出缓冲区。

原图如图 10-1 所示。

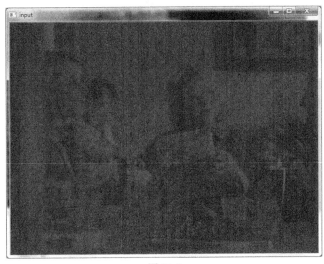

图 10-1

如果设置为"bool useLogSampling = true;"时，执行结果如图 10-2 所示。

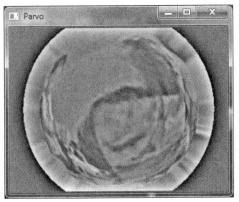 

图 10-2

如果设成"bool useLogSampling = false;"时，执行结果如图 10-3 所示。

图 10-3

如果想要查看视频效果，就将程序改为 string inputMediaType = "-video"。

第11章

实际应用

11.1 图像藏密

密码学的应用流行多年并且技巧繁多。本节所要介绍的是图像藏密（image steganography）的隐藏技术。而密码学分为加密与解密，在此分别以两个程序说明，先介绍加密再介绍解密。

```cpp
#include <opencv2/core/core.hpp>
#include <opencv2/highgui/highgui.hpp>
#include <iostream>
using namespace cv;
using namespace std;
int main(int argc, char** argv)
{
    Mat image1, image2, image3;
    image1 = imread("C:\\images\\lena.jpg", IMREAD_COLOR);
    image2 = imread("C:\\images\\baboon.jpg", IMREAD_COLOR);
    namedWindow("Org image", WINDOW_AUTOSIZE);
    namedWindow("Hide image", WINDOW_AUTOSIZE);
    namedWindow("Steged image", WINDOW_AUTOSIZE);
    // 检查两个图的大小与类型
    if (image1.type() != image2.type() ||
        image1.size() != image2.size())
    {
        printf("两图类型或大小不同   \n");
        return 1;
    }
    // 图像的高；行数
    int numberRows = image1.rows;
    // 图像的宽；列数
    int numberCols = image1.cols;
    // 产生加密文件
    image3.create(numberRows, numberCols, image1.type());
    Mat tFront_image, tHidden_image;
    // 要显示的图像
    Mat front_mask(numberRows, numberCols, image1.type(),
        Scalar(0xF0, 0xF0, 0xF0));
    // 要隐藏的图像
    Mat hidden_mask(numberRows, numberCols, image1.type(),
     Scalar(0xF0, 0xF0, 0xF0));
    // 前两图进行位的相加(and)处理，结果放入第三张图
    // 因为之前声明资料为 0xF0
    // 相加之后只保留前四个位
    bitwise_and(image1, front_mask, tFront_image);
    bitwise_and(image2, hidden_mask, tHidden_image);
    // 处理每个颜色通道，将左侧四个位移到右侧
    for (int j = 0; j < numberRows; j++)
        for (int i = 0; i < numberCols; i++){
            tHidden_image.at<Vec3b>(j, i)[0] = tHidden_image.
                at<Vec3b>(j, i)[0] >> 4;
            tHidden_image.at<Vec3b>(j, i)[1] = tHidden_image.
                at<Vec3b>(j, i)[1] >> 4;
```

```
                tHidden_image.at<Vec3b>(j, i)[2] = tHidden_image.
                    at<Vec3b>(j, i)[2] >> 4;
        }
    // 前两图进行位的互补(or)处理，结果放入第三张图
    // 要隐藏的图就是右侧四个位
    bitwise_or(tFront_image, tHidden_image, image3);
    imshow("Org image", image1);
    imshow("Hide image", image2);
    imshow("Steged image", image3);
    imwrite("C:\\images\\process\\staged-lena.jpg", image3);
    waitKey(0);
    return 0;
}
```

1. bitwise_and(InputArray src1, InputArray src2, OutputArray dst, InputArray mask=noArray())：计算两个图像内每个元素（per-element）位的连接（conjuction）

（1）src1：第一输入图像或 Scalar 颜色值。

（2）src2：第二输入图像或 Scalar 颜色值。

（3）dst：输出图像，与输入图像同大小与类型。

（4）mask：可有可无的掩码。

2. bitwise_or(InputArray src1, InputArray src2, OutputArray dst, InputArray mask=noArray())：计算两个图像内每个元素位的分离（disjuction）

（1）src1：第一输入图像或 Scalar 颜色值。

（2）src2：第二输入图像或 Scalar 颜色值。

（3）dst：输出图像，与输入图像同大小与类型。

（4）mask：是可有可无的掩码。

执行结果如图 11-1 所示。

（a）原图　　　　　　　（b）要隐藏的图　　　　　　　（c）原图加隐藏图

图 11-1

程序中的加密原则，是认为每个字节（byte）的各个位都有其重要性。而重要性是从左到右降序，左侧是重要的位，称为最高有效位（Most Significant Bit，MSB）；右侧是不重要的位，称为最低有效位（Least Significant Bit，LSB）。所以本程序要将隐藏的重要的位，放到另一个字节的最低有效位。

本程序只是示范，所以加密前后两个文件的大小与图文件的类型都必须相同。例如，使用同一台相机或手机拍摄的图像大小一定是相同的，除了手机横拍或直拍的差异。不过相信读者已知道要被隐藏的图像其长宽一定要比较小，因为在两层的 for loop 处理的技巧上，超过隐藏文件的长或宽就不进行处理了。

图像解密

图像解密的代码如下：

```cpp
#include <stdio.h>
#include "opencv2/highgui/highgui.hpp"
#include "opencv2/core/core.hpp"
#include "opencv2/imgproc/imgproc.hpp"
using namespace cv;
using namespace std;
int main(int argc, char** argv)
{
    Mat image2, image3;
    image3 = imread("C:\\images\\process\\staged-lena.jpg");
    // 图像的高；行数
    int numberRows = image3.rows;
    // 图像的宽；列数
    int numberCols = image3.cols;
    // 产生解密文件
    image2.create(numberRows, numberCols, image3.type());
    Mat tHidden_image;
    Mat hidden_mask(numberRows, numberCols,
        image3.type(), Scalar(0x0F, 0x0F, 0x0F));
    bitwise_and(image3, hidden_mask, image2);
    // 还原加密处理
    for (int j = 0; j < numberRows; j++)
        for (int i = 0; i < numberCols; i++)
        {
            image2.at<Vec3b>(j, i)[0] =
                image2.at<Vec3b>(j, i)[0] << 4;
            image2.at<Vec3b>(j, i)[1] =
                image2.at<Vec3b>(j, i)[1] << 4;
            image2.at<Vec3b>(j, i)[2] =
                image2.at<Vec3b>(j, i)[2] << 4;
        }
    imshow("Staged Image", image3);
    imshow("Hidden Image", image2);
    waitKey(0);
    return 0;
}
```

执行结果如图 11-2 所示。

（a）原图加隐藏图　　　　　　　　　　　（b）解密出的图像

图 11-2

也许读者认为图片有失真，其实隐藏图像并不一定是要传送真实的图片，而只是为了传递图像中的信息。

11.2　图像采集

一般来说，进行图像处理并不是要处理整个图像，而是希望针对原图的特定的区域进行处理。此时就需要先用鼠标采集我们期望处理的区域，采集后再进行图像处理，所以本节示范如何采集图像区域。

采集图像区域的代码如下：

```
#include <opencv2/core/core.hpp>
#include <opencv2/highgui/highgui.hpp>
#include <opencv2/imgproc/imgproc.hpp>
using namespace cv;

// 声明函数
void my_mouse_callback(int event, int x, int y,
    int flags, void* param);
// 声明全局变量
Mat image;
Rect box;
bool drawing_box = false;

int main( int argc, char* argv[] )
{
    image = imread("c:/images/lena.jpg", 1);
    namedWindow( "Box Example", CV_WINDOW_AUTOSIZE);
    imshow("Box Example", image);
    setMouseCallback("Box Example", my_mouse_callback);
```

```cpp
    while( 1 )
    {
        if ( waitKey(15)==27 )
            break;
    }
    return 0;
}

// 鼠标回调函数
void my_mouse_callback(int event, int x, int y,
    int flags, void* parm)
{
    switch( event )
    {
        // 移动鼠标
        case EVENT_MOUSEMOVE:
        {
            if( drawing_box )
            {
                box.width  = x-box.x;
                box.height = y-box.y;
                Mat temp = image.clone();
                // 绘制选取内容
                rectangle(temp, Point(box.x, box.y),
                    Point(box.x + box.width, box.y + box.height),
                    Scalar(0, 0xff, 0), 3);
                imshow("Box Example", temp);
            }
        }
        break;
        // 按下鼠标左键
        case EVENT_LBUTTONDOWN:
        {
            drawing_box = true;
            box = Rect( x, y, 0, 0 );
        }
        break;
        // 释放鼠标左键
        case EVENT_LBUTTONUP:
        {
            drawing_box = false;
            if( box.width<0 )
            {
                box.x+=box.width;
                box.width *=-1;
            }
            if( box.height<0 )
            {
                box.y+=box.height;
                box.height*=-1;
            }
            // 采集图像
            Mat temp(image, box);
            imshow("Crop", temp);
            imwrite("c:/images/process/temp.jpg", temp);
            // 绘制选取内容
```

```
            rectangle(image, Point(box.x, box.y),
                Point(box.x+box.width, box.y+box.height),
                Scalar(0, 0xff, 0), 3);
            imshow("Box Example", image);
        }
        break;
    }
}
```

1. **setMouseCallback(const string& winname, MouseCallback onMouse, void* userdata=0)**：设置鼠标的回调函数

    （1）winname：窗口名称。

    （2）onMouse：回调实际处理函数。

    （3）userdata：传给回调处理函数的参数，是可有可无的参数。

2. **mouse_callback(par1, par2, par3, par4, par5)**：鼠标的回调函数

    par1 是发生的事件，事件有下列几种类型。

类型名称	数值	操作
CV_EVENT_MOUSEMOVE	0	鼠标移动
CV_EVENT_LBUTTONDOWN	1	按下鼠标左键
CV_EVENT_RBUTTONDOWN	2	按下鼠标右键
CV_EVENT_MBUTTONDOWN	3	按下鼠标中键
CV_EVENT_LBUTTONUP	4	释放鼠标左键
CV_EVENT_RBUTTONUP	5	释放鼠标右键
CV_EVENT_MBUTTONUP	6	释放鼠标中键
CV_EVENT_LBUTTONDBLCLK	7	双点鼠标左键
CV_EVENT_RBUTTONDBLCLK	8	双点鼠标右键
CV_EVENT_MBUTTONDBLCLK	9	双点鼠标中键

   par2 与 par3 是鼠标坐标。

   par4 是事件发生时的特殊条件。

类型名称	数值	操作
CV_EVENT_FLAG_LBUTTON	1	按住鼠标左键
CV_EVENT_FLAG_RBUTTON	2	按住鼠标右键
CV_EVENT_FLAG_MBUTTON	4	按住鼠标中键
CV_EVENT_FLAG_CTRLKEY	8	按下 Ctrl 键
CV_EVENT_FLAG_SHIFTKEY	16	按下 Shift 键
CV_EVENT_FLAG_ALTKEY	32	按下 Alt 键

执行结果如图 11-3 所示。

（a）原图　　　　　　　　　　　（b）执行采集中

图 11-3

键盘应用

键盘应用的代码如下：

```cpp
#include "opencv2/highgui/highgui.hpp"
#include "opencv2/imgproc/imgproc.hpp"
#include <iostream>
using namespace std;
using namespace cv;
static void help()
{
    cout << "\n热键: \n"
        "\tESC - 结束程序\n"
        "\tr - 还原图像\n"
        "\tn - 开始操作\n"
        "\n"
        "\t鼠标左键 - 设置区域\n"
        "\n"
        "\tCTRL+鼠标左键 - set GC_BGD pixels\n"
        "\tSHIFT+鼠标左键 - set CG_FGD pixels\n"
        "\n"
        "\tCTRL+鼠标右键 - set GC_PR_BGD pixels\n"
        "\tSHIFT+鼠标右键 - set CG_PR_FGD pixels\n" << endl;
}
const Scalar RED = Scalar(0,0,255);
const Scalar PINK = Scalar(230,130,255);
const Scalar BLUE = Scalar(255,0,0);
const Scalar LIGHTBLUE = Scalar(255,255,160);
const Scalar GREEN = Scalar(0,255,0);
const int BGD_KEY = CV_EVENT_FLAG_CTRLKEY;
const int FGD_KEY = CV_EVENT_FLAG_SHIFTKEY;
static void getBinMask( const Mat& comMask, Mat& binMask )
{
    if( comMask.empty() || comMask.type()!=CV_8UC1 )
```

```cpp
            CV_Error( CV_StsBadArg,
          "comMask is empty or has incorrect type (not CV_8UC1)" );
    if( binMask.empty() || binMask.rows!=comMask.rows
            || binMask.cols!=comMask.cols )
        binMask.create( comMask.size(), CV_8UC1 );
    binMask = comMask & 1;
}
class GCApplication
{
public:
    enum{ NOT_SET = 0, IN_PROCESS = 1, SET = 2 };
    static const int radius = 2;
    static const int thickness = -1;
    void reset();
    void setImageAndWinName( const Mat& _image,
        const string& _winName );
    void showImage() const;
    void mouseClick( int event, int x, int y,
        int flags, void* param );
    int nextIter();
    int getIterCount() const { return iterCount; }
private:
    void setRectInMask();
    void setLblsInMask( int flags, Point p, bool isPr );
    const string* winName;
    const Mat* image;
    Mat mask;
    Mat bgdModel, fgdModel;
    uchar rectState, lblsState, prLblsState;
    bool isInitialized;
    Rect rect;
    vector<Point> fgdPxls, bgdPxls, prFgdPxls, prBgdPxls;
    int iterCount;
};
void GCApplication::reset()
{
    if( !mask.empty() )
        mask.setTo(Scalar::all(GC_BGD));
    bgdPxls.clear();   fgdPxls.clear();
    prBgdPxls.clear(); prFgdPxls.clear();
    isInitialized = false;
    rectState = NOT_SET;
    lblsState = NOT_SET;
    prLblsState = NOT_SET;
    iterCount = 0;
}
void GCApplication::setImageAndWinName( const Mat& _image,
    const string& _winName)
{
    if( _image.empty() || _winName.empty() )
        return;
    image = &_image;
    winName = &_winName;
    mask.create( image->size(), CV_8UC1);
    reset();
}
void GCApplication::showImage() const
```

```cpp
{
    if( image->empty() || winName->empty() )
        return;
    Mat res;
    Mat binMask;
    if( !isInitialized )
        image->copyTo( res );
    else
    {
        getBinMask( mask, binMask );
        image->copyTo( res, binMask );
    }
    vector<Point>::const_iterator it;
    for( it = bgdPxls.begin(); it != bgdPxls.end(); ++it )
        circle( res, *it, radius, BLUE, thickness );
    for( it = fgdPxls.begin(); it != fgdPxls.end(); ++it )
        circle( res, *it, radius, RED, thickness );
    for( it = prBgdPxls.begin(); it != prBgdPxls.end(); ++it )
        circle( res, *it, radius, LIGHTBLUE, thickness );
    for( it = prFgdPxls.begin(); it != prFgdPxls.end(); ++it )
        circle( res, *it, radius, PINK, thickness );
    if( rectState == IN_PROCESS || rectState == SET )
        rectangle( res, Point( rect.x, rect.y ),
            Point(rect.x + rect.width, rect.y + rect.height ),
            GREEN, 2);
    imshow( *winName, res );
}
void GCApplication::setRectInMask()
{
    assert( !mask.empty() );
    mask.setTo( GC_BGD );
    rect.x = max(0, rect.x);
    rect.y = max(0, rect.y);
    rect.width = min(rect.width, image->cols-rect.x);
    rect.height = min(rect.height, image->rows-rect.y);
    (mask(rect)).setTo( Scalar(GC_PR_FGD) );
}
void GCApplication::setLblsInMask( int flags, Point p, bool isPr )
{
    vector<Point> *bpxls, *fpxls;
    uchar bvalue, fvalue;
    if( !isPr )
    {
        bpxls = &bgdPxls;
        fpxls = &fgdPxls;
        bvalue = GC_BGD;
        fvalue = GC_FGD;
    }
    else
    {
        bpxls = &prBgdPxls;
        fpxls = &prFgdPxls;
        bvalue = GC_PR_BGD;
        fvalue = GC_PR_FGD;
    }
    if( flags & BGD_KEY )
```

```cpp
    {
        bpxls->push_back(p);
        circle( mask, p, radius, bvalue, thickness );
    }
    if( flags & FGD_KEY )
    {
        fpxls->push_back(p);
        circle( mask, p, radius, fvalue, thickness );
    }
}
void GCApplication::mouseClick( int event, int x,
    int y, int flags, void* )
{
    // TODO add bad args check
    switch( event )
    {
    // set rect or GC_BGD(GC_FGD) labels
    case CV_EVENT_LBUTTONDOWN:
        {
            bool isb = (flags & BGD_KEY) != 0,
                isf = (flags & FGD_KEY) != 0;
            if( rectState == NOT_SET && !isb && !isf )
            {
                rectState = IN_PROCESS;
                rect = Rect( x, y, 1, 1 );
            }
            if ( (isb || isf) && rectState == SET )
                lblsState = IN_PROCESS;
        }
        break;
    // set GC_PR_BGD(GC_PR_FGD) labels
    case CV_EVENT_RBUTTONDOWN:
        {
            bool isb = (flags & BGD_KEY) != 0,
                isf = (flags & FGD_KEY) != 0;
            if ( (isb || isf) && rectState == SET )
                prLblsState = IN_PROCESS;
        }
        break;
    case CV_EVENT_LBUTTONUP:
        if( rectState == IN_PROCESS )
        {
            rect = Rect( Point(rect.x, rect.y), Point(x,y) );
            rectState = SET;
            setRectInMask();
            assert( bgdPxls.empty() && fgdPxls.empty() &&
                prBgdPxls.empty() && prFgdPxls.empty() );
            showImage();
        }
        if( lblsState == IN_PROCESS )
        {
            setLblsInMask(flags, Point(x,y), false);
            lblsState = SET;
            showImage();
        }
```

```cpp
            break;
        case CV_EVENT_RBUTTONUP:
            if( prLblsState == IN_PROCESS )
            {
                setLblsInMask(flags, Point(x,y), true);
                prLblsState = SET;
                showImage();
            }
            break;
        case CV_EVENT_MOUSEMOVE:
            if( rectState == IN_PROCESS )
            {
                rect = Rect( Point(rect.x, rect.y), Point(x,y) );
                assert( bgdPxls.empty() && fgdPxls.empty()
                    && prBgdPxls.empty() && prFgdPxls.empty() );
                showImage();
            }
            else if( lblsState == IN_PROCESS )
            {
                setLblsInMask(flags, Point(x,y), false);
                showImage();
            }
            else if( prLblsState == IN_PROCESS )
            {
                setLblsInMask(flags, Point(x,y), true);
                showImage();
            }
            break;
    }
}
int GCApplication::nextIter()
{
    if( isInitialized )
        grabCut( *image, mask, rect, bgdModel, fgdModel, 1 );
    else
    {
        if( rectState != SET )
            return iterCount;
        if( lblsState == SET || prLblsState == SET )
            grabCut( *image, mask, rect, bgdModel,
                fgdModel, 1, GC_INIT_WITH_MASK );
        else
            grabCut( *image, mask, rect, bgdModel,
                fgdModel, 1, GC_INIT_WITH_RECT );
        isInitialized = true;
    }
    iterCount++;
    bgdPxls.clear(); fgdPxls.clear();
    prBgdPxls.clear(); prFgdPxls.clear();
    return iterCount;
}
GCApplication gcapp;
static void on_mouse( int event, int x, int y, int flags, void* param )
{
    gcapp.mouseClick( event, x, y, flags, param );
```

```cpp
}
int main( int argc, char** argv )
{
    argv[1] = "c:/images/lena.jpg";
    string filename = argv[1];
    if( filename.empty() )
    {
        cout << "\n无效文件名 " << argv[1] << endl;
        return 1;
    }
    Mat image = imread( filename, 1 );
    if( image.empty() )
    {
        cout << "\n无法读取图文件 " << filename << endl;
        return 1;
    }
    help();
    const string winName = "image";
    namedWindow( winName, WINDOW_AUTOSIZE );
    setMouseCallback( winName, on_mouse, 0 );
    gcapp.setImageAndWinName( image, winName );
    gcapp.showImage();
    for(;;)
    {
        int c = waitKey(0);
        switch( (char) c )
        {
        case '\x1b':
            cout << "Exiting ..." << endl;
            goto exit_main;
        case 'r':
            cout << endl;
            gcapp.reset();
            gcapp.showImage();
            break;
        case 'n':
            int iterCount = gcapp.getIterCount();
            cout << "<" << iterCount << "... ";
            int newIterCount = gcapp.nextIter();
            if( newIterCount > iterCount )
            {
                gcapp.showImage();
                cout << iterCount << ">" << endl;
            }
            else
                cout << "请用鼠标选取区域" << endl;
            break;
        }
    }
exit_main:
    destroyWindow( winName );
    return 0;
}
```

执行结果如图 11-4 所示。

图 11-4

图像分割

图像分割的代码如下:

```
#include "opencv2/highgui/highgui.hpp"
#include "opencv2/core/core.hpp"
using namespace std;
using namespace cv;
int main()
{
   Mat img = imread("C:\\images\\lena.jpg");
   imshow("original", img);
   for (int i = 0; i <= img.rows - (img.rows / 2);
      i = i + (img.rows / 2))
   {
      for (int j = 0; j <= img.cols - (img.cols / 2);
         j = j + (img.cols / 2))
      {
         Mat sub(img, Rect(j, i, (img.rows / 2), (img.rows / 2)));
         imwrite("C:\\images\\process\\sub"
            + to_string(static_cast<long long>(i))
            + to_string(static_cast<long long>(j))
            + ".jpg", sub);
      }
   }
   waitKey(0);
   return 0;
}
```

执行结果如图 11-5 所示。

图 11-5

图像翻转

图像翻转的代码如下：

```cpp
#include <iostream>
#include <opencv2/core/core.hpp>
#include <opencv2/highgui/highgui.hpp>
using namespace cv;
int main() {
   Mat image;
   image= imread("C:\\images\\lena.jpg");
   if (!image.data)
      return 0;
   // 显示原图
   namedWindow("Original Image");
    imshow("Original Image", image);
   Mat result;
   // 翻转图像
   // 左右翻转
   flip(image, result, 1);
   // 显示翻转结果
   namedWindow("Output Image");
   imshow("Output Image", result);
   // 上下翻转
   flip(image, result, 0);
   //显示翻转结果
   namedWindow("image 2");
   imshow("image 2", result);
   // 上下+左右翻转
   flip(image, result, -1);
   namedWindow("image 3");
   imshow("image 3", result);
   waitKey(0);
   return 0;
}
```

flip(InputArray src, OutputArray dst, int flipCode)：二维图像垂直、水平或两者都翻转

（1）src：输入图像。

（2）dst：输出图像。

（3）flipCode：用来指定如何翻转的标志。

执行结果如图11-6所示。

图 11-6

鼠标坐标检测

鼠标坐标检测的代码如下：

```
#include <iostream>
#include <opencv2/highgui/highgui.hpp>
#include <opencv2/imgproc/imgproc.hpp>
using namespace std;
using namespace cv;
// 声明全局变量
// 鼠标左键的判定值
bool ldown = false, lup = false;
// 图文件
Mat img;
// 字体
```

```
int fontFace = FONT_HERSHEY_SIMPLEX;
// 字体大小
double fontScale = 0.3;
// 文字笔划粗细度
int thickness = 0.5;
char text[10];
// 鼠标事件的回调函数
static void mouse_callback(int event, int x, int y, int, void *)
{
    // 鼠标左键按下
    if (event == EVENT_LBUTTONDOWN)
    {
        Mat local_img = img.clone();
        sprintf(text, "(%d, %d)", x, y);
        putText(local_img, text, Point(x, y), fontFace, fontScale,
            Scalar::all(255), thickness, 1);

        imshow("鼠标坐标", local_img);
    }
}
int main()
{
    img = imread("C:\\images\\lena.jpg");
    namedWindow("鼠标坐标", CV_WINDOW_AUTOSIZE);
    imshow("鼠标坐标", img);
    // 鼠标事件的回调函数
    setMouseCallback("鼠标坐标", mouse_callback);
    // 按 'q' 结束
    while (char(waitKey(1)) != 'q') {}
    return 0;
}
```

Mat Mat::clone() const：图像复制，如图 11-7 所示。

因为图形处理往往要知道处理的区域(ROI)，通过取得鼠标坐标点就方便找到 ROI 坐标，例如，在 2.3 节中，我们就是以此方式找到商标该放置的位置。

图 11-7

11.3 QR Code 检测

QR Code 检测的代码如下:

```cpp
#include <opencv2/core/core.hpp>
#include <opencv2/highgui/highgui.hpp>
#include "opencv2/imgproc/imgproc.hpp"
#include <iostream>
#include <cmath>
using namespace cv;
using namespace std;
const int CV_QR_NORTH = 0;
const int CV_QR_EAST = 1;
const int CV_QR_SOUTH = 2;
const int CV_QR_WEST = 3;
float cv_distance(Point2f P, Point2f Q);
float cv_lineEquation(Point2f L, Point2f M, Point2f J);
float cv_lineSlope(Point2f L, Point2f M, int& alignement);
void cv_getVertices(vector<vector<Point> > contours,
    int c_id,float slope, vector<Point2f>& X);
void cv_updateCorner(Point2f P, Point2f ref,
    float& baseline,  Point2f& corner);
void cv_updateCornerOr(int orientation,
    vector<Point2f> IN, vector<Point2f> &OUT);
bool getIntersectionPoint(Point2f a1, Point2f a2,
    Point2f b1, Point2f b2, Point2f& intersection);
float cross(Point2f v1,Point2f v2);
int main ( int argc, char **argv )
{
   Mat image = imread("c:/images/QRCode1.jpg");
   if (image.empty())
   {
      cerr << "图文件读取失败\n" << endl;
      return -1;
   }
   Mat gray(image.size(), CV_MAKETYPE(image.depth(), 1));
   Mat edges(image.size(), CV_MAKETYPE(image.depth(), 1));
   Mat qr, qr_raw, qr_gray, qr_thres;
   cvtColor(image, gray, CV_RGB2GRAY);

   vector<vector<Point> > contours;
   vector<Vec4i> hierarchy;
   int mark,A,B,C,top,right,bottom,median1,median2,outlier;
   float AB,BC,CA, dist,slope, areat,arear,areab, large, padding;

   int align,orientation;
   qr_raw = Mat::zeros(100, 100, CV_8UC3 );
   qr = Mat::zeros(100, 100, CV_8UC3 );
   qr_gray = Mat::zeros(100, 100, CV_8UC1 );
   qr_thres = Mat::zeros(100, 100, CV_8UC1 );

   Canny(gray, edges, 100 , 200, 3);
   findContours( edges, contours, hierarchy,
      RETR_TREE, CHAIN_APPROX_SIMPLE);
```

```
mark = 0;
vector<Moments> mu(contours.size());
  vector<Point2f> mc(contours.size());
for( int i = 0; i < contours.size(); i++ )
{
   mu[i] = moments( contours[i], false );
   mc[i] = Point2f( mu[i].m10/mu[i].m00 ,
      mu[i].m01/mu[i].m00 );
}

for( int i = 0; i < contours.size(); i++ )
{
   int k=i;
   int c=0;
   while(hierarchy[k][2] != -1)
   {
      k = hierarchy[k][2] ;
      c = c+1;
   }
   if(hierarchy[k][2] != -1)
      c = c+1;
   if (c >= 5)
   {
      if (mark == 0)  A = i;
      else if  (mark == 1) B = i;
      else if  (mark == 2) C = i;
      mark = mark + 1 ;
   }
}

if (mark >= 2)
{
   AB = cv_distance(mc[A],mc[B]);
   BC = cv_distance(mc[B],mc[C]);
   CA = cv_distance(mc[C],mc[A]);

   if ( AB > BC && AB > CA )
   {
      outlier = C; median1=A; median2=B;
   }
   else if ( CA > AB && CA > BC )
   {
      outlier = B; median1=A; median2=C;
   }
   else if ( BC > AB && BC > CA )
   {
      outlier = A;  median1=B; median2=C;
   }

   top = outlier;

   dist = cv_lineEquation(mc[median1],
      mc[median2], mc[outlier]);
   slope = cv_lineSlope(mc[median1], mc[median2],align);
   if (align == 0)
```

```
        {
            bottom = median1;
            right = median2;
        }
        else if (slope < 0 && dist < 0 )
        {
            bottom = median1;
            right = median2;
            orientation = CV_QR_NORTH;
        }
        else if (slope > 0 && dist < 0 )
        {
            right = median1;
            bottom = median2;
            orientation = CV_QR_EAST;
        }
        else if (slope < 0 && dist > 0 )
        {
            right = median1;
            bottom = median2;
            orientation = CV_QR_SOUTH;
        }
        else if (slope > 0 && dist > 0 )
        {
            bottom = median1;
            right = median2;
            orientation = CV_QR_WEST;
        }

        float area_top,area_right, area_bottom;

        if( top < contours.size() &&
            right < contours.size() &&
            bottom < contours.size() &&
            contourArea(contours[top]) > 10 &&
            contourArea(contours[right]) > 10 &&
            contourArea(contours[bottom]) > 10 )
        {
            vector<Point2f> L,M,O, tempL,tempM,tempO;
            Point2f N;
            vector<Point2f> src,dst;
            Mat warp_matrix;
            cv_getVertices(contours,top,slope,tempL);
            cv_getVertices(contours,right,slope,tempM);
            cv_getVertices(contours,bottom,slope,tempO);
            cv_updateCornerOr(orientation, tempL, L);
            cv_updateCornerOr(orientation, tempM, M);
            cv_updateCornerOr(orientation, tempO, O);
            int iflag =
                    getIntersectionPoint(M[1],M[2],O[3],O[2],N);
            src.push_back(L[0]);
            src.push_back(M[1]);
            src.push_back(N);
            src.push_back(O[3]);
```

```cpp
            dst.push_back(Point2f(0,0));
            dst.push_back(Point2f(qr.cols,0));
            dst.push_back(Point2f(qr.cols, qr.rows));
            dst.push_back(Point2f(0, qr.rows));
            if (src.size() == 4 && dst.size() == 4 )
            {
                warp_matrix = getPerspectiveTransform(src, dst);
                warpPerspective(image, qr_raw, warp_matrix,
                    Size(qr.cols, qr.rows));
                copyMakeBorder( qr_raw, qr, 10, 10, 10, 10,
                    BORDER_CONSTANT, Scalar(255,255,255) );

                cvtColor(qr,qr_gray,CV_RGB2GRAY);
                threshold(qr_gray, qr_thres, 127,
                    255, CV_THRESH_BINARY);
            }

            drawContours( image, contours, top,
                Scalar(255,200,0), 2, 8, hierarchy, 0 );
            drawContours( image, contours, right,
                Scalar(0,0,255), 2, 8, hierarchy, 0 );
            drawContours( image, contours, bottom,
                Scalar(255,0,100), 2, 8, hierarchy, 0 );
        }
    }
    namedWindow("Image", CV_WINDOW_NORMAL);

    imshow ( "Image", image );
    imshow ( "QR code", qr_thres );
    waitKey(0);
    return 0;
}
float cv_distance(Point2f P, Point2f Q)
{
    return sqrt(pow(abs(P.x - Q.x),2) + pow(abs(P.y - Q.y),2)) ;
}
float cv_lineEquation(Point2f L, Point2f M, Point2f J)
{
    float a,b,c,pdist;
    a = -((M.y - L.y) / (M.x - L.x));
    b = 1.0;
    c = (((M.y - L.y) /(M.x - L.x)) * L.x) - L.y;
    pdist = (a * J.x + (b * J.y) + c) / sqrt((a * a) + (b * b));
    return pdist;
}
float cv_lineSlope(Point2f L, Point2f M, int& alignement)
{
    float dx,dy;
    dx = M.x - L.x;
    dy = M.y - L.y;

    if ( dy != 0)
    {
        alignement = 1;
        return (dy / dx);
```

```cpp
        }
        else
        {
            alignement = 0;
            return 0.0;
        }
}
void cv_getVertices(vector<vector<Point> > contours,
    int c_id, float slope, vector<Point2f>& quad)
{
    Rect box;
    box = boundingRect( contours[c_id]);

    Point2f M0,M1,M2,M3;
    Point2f A, B, C, D, W, X, Y, Z;
    A =  box.tl();
    B.x = box.br().x;
    B.y = box.tl().y;
    C = box.br();
    D.x = box.tl().x;
    D.y = box.br().y;
    W.x = (A.x + B.x) / 2;
    W.y = A.y;
    X.x = B.x;
    X.y = (B.y + C.y) / 2;
    Y.x = (C.x + D.x) / 2;
    Y.y = C.y;
    Z.x = D.x;
    Z.y = (D.y + A.y) / 2;
    float dmax[4];
    dmax[0]=0.0;
    dmax[1]=0.0;
    dmax[2]=0.0;
    dmax[3]=0.0;
    float pd1 = 0.0;
    float pd2 = 0.0;
    if (slope > 5 || slope < -5 )
    {
        for( int i = 0; i < contours[c_id].size(); i++ )
        {
            pd1 = cv_lineEquation(C, A, contours[c_id][i]);
            pd2 = cv_lineEquation(B,D,contours[c_id][i]);
            if((pd1 >= 0.0) && (pd2 > 0.0))
            {
                cv_updateCorner(contours[c_id][i],W,dmax[1],M1);
            }
            else if((pd1 > 0.0) && (pd2 <= 0.0))
            {
                cv_updateCorner(contours[c_id][i],X,dmax[2],M2);
            }
            else if((pd1 <= 0.0) && (pd2 < 0.0))
            {
                cv_updateCorner(contours[c_id][i],Y,dmax[3],M3);
            }
            else if((pd1 < 0.0) && (pd2 >= 0.0))
```

```
            {
                cv_updateCorner(contours[c_id][i],Z,dmax[0],M0);
            }
            else
                continue;
            }
        }
        else
        {
            int halfx = (A.x + B.x) / 2;
            int halfy = (A.y + D.y) / 2;
            for( int i = 0; i < contours[c_id].size(); i++ )
            {
                if((contours[c_id][i].x < halfx) &&
                    (contours[c_id][i].y <= halfy))
                {
                    cv_updateCorner(contours[c_id][i],C,dmax[2],M0);
                }
                else if((contours[c_id][i].x >= halfx) &&
                    (contours[c_id][i].y < halfy))
                {
                    cv_updateCorner(contours[c_id][i],D,dmax[3],M1);
                }
                else if((contours[c_id][i].x > halfx) &&
                    (contours[c_id][i].y >= halfy))
                {
                    cv_updateCorner(contours[c_id][i],A,dmax[0],M2);
                }
                else if((contours[c_id][i].x <= halfx) &&
                    (contours[c_id][i].y > halfy))
                {
                    cv_updateCorner(contours[c_id][i],B,dmax[1],M3);
                }
            }
        }
    quad.push_back(M0);
    quad.push_back(M1);
    quad.push_back(M2);
    quad.push_back(M3);
}
void cv_updateCorner(Point2f P, Point2f ref ,
    float& baseline,   Point2f& corner)
{
    float temp_dist;
    temp_dist = cv_distance(P,ref);
    if(temp_dist > baseline)
    {
        baseline = temp_dist;
        corner = P;
    }

}
void cv_updateCornerOr(int orientation,
    vector<Point2f> IN,vector<Point2f> &OUT)
{
```

```
        Point2f M0,M1,M2,M3;
        if(orientation == CV_QR_NORTH)
    {
        M0 = IN[0];
        M1 = IN[1];
        M2 = IN[2];
        M3 = IN[3];
    }
    else if (orientation == CV_QR_EAST)
    {
        M0 = IN[1];
        M1 = IN[2];
        M2 = IN[3];
        M3 = IN[0];
    }
    else if (orientation == CV_QR_SOUTH)
    {
        M0 = IN[2];
        M1 = IN[3];
        M2 = IN[0];
        M3 = IN[1];
    }
    else if (orientation == CV_QR_WEST)
    {
        M0 = IN[3];
        M1 = IN[0];
        M2 = IN[1];
        M3 = IN[2];
    }
    OUT.push_back(M0);
    OUT.push_back(M1);
    OUT.push_back(M2);
    OUT.push_back(M3);
}
bool getIntersectionPoint(Point2f a1, Point2f a2,
    Point2f b1, Point2f b2, Point2f& intersection)
{
    Point2f p = a1;
    Point2f q = b1;
    Point2f r(a2-a1);
    Point2f s(b2-b1);
    if(cross(r,s) == 0) {return false;}
    float t = cross(q-p,s)/cross(r,s);
    intersection = p + t*r;
    return true;
}
float cross(Point2f v1,Point2f v2)
{
    return v1.x*v2.y - v1.y*v2.x;
}
```

执行结果如图 11-8 所示。

（a）原图　　　　　　　　　　（b）检测结果

图 11-8

11.4　与 OpenGL 整合

与 OpenGL 整合的代码如下：

```
#include <glut.h>
#include <Gl/gl.h>
#include "glext.h"
#include <opencv2/core/core.hpp>
#include <opencv2/highgui/highgui.hpp>
using namespace cv;
// 纹理的矩阵
GLuint texture;
// 旋转角度
GLfloat angle = 0.0;
// 传送 image 给 OpenGL 纹理函数
int loadTexture(Mat image, GLuint *text)
{
    if (image.empty())
        return -1;

    // 建立 OpenGL 纹理
    glGenTextures(1, text);
    // 绑定纹理到阵列
    glBindTexture( GL_TEXTURE_2D, *text );
    // 设置纹理参数
    glTexParameteri(GL_TEXTURE_2D,
        GL_TEXTURE_MAG_FILTER,GL_LINEAR);
    glTexParameteri(GL_TEXTURE_2D,
        GL_TEXTURE_MIN_FILTER,GL_LINEAR);

    // 设置像素存储模式
    glPixelStorei(GL_UNPACK_ALIGNMENT, 1);
    // 指定二维纹理图像
```

```
    glTexImage2D(GL_TEXTURE_2D, 0, GL_RGB, image.cols,
        image.rows, 0, GL_BGR, GL_UNSIGNED_BYTE, image.data);
    return 0;
}
void plane (void)
{
    // 绑定纹理
    glBindTexture( GL_TEXTURE_2D, texture );
    // 以旋转矩阵相乘目前矩阵
    glRotatef( angle, 1.0f, 1.0f, 1.0f );
    glBegin (GL_QUADS);

    // 纹理坐标的顶点(vertices)
    glTexCoord2d(0.0, 0.0);
    glVertex2d(-1.0, -1.0);

    // 设置纹理绘图点
    glTexCoord2d(1.0, 0.0); // 纹理坐标
    glVertex2d(+1.0, -1.0); // 指定顶点(Vertex)
    glTexCoord2d(1.0, 1.0);
    glVertex2d(+1.0, +1.0);
    glTexCoord2d(0.0, 1.0);
    glVertex2d(-1.0, +1.0);
    glEnd();

}
void display (void)
{
    // 清除颜色缓冲区的值
    glClearColor (0.0,0.0,0.0,1.0);
    // 以目前值清除缓冲区
    glClear (GL_COLOR_BUFFER_BIT);
    // 以相同矩阵取代现有矩阵
    glLoadIdentity();
    // 定义视角转换
    gluLookAt (0.0, 0.0, 5.0, 0.0, 0.0, 0.0, 0.0, 1.0, 0.0);

    // 启用 2D 纹理
    glEnable( GL_TEXTURE_2D );
        plane();
    // 缓冲区互换
    glutSwapBuffers();
    angle =angle+0.005;
}
void FreeTexture( GLuint texture )
{
    // 删除纹理
    glDeleteTextures( 1, &texture );
}
void reshape (int w, int h)
{
    // 设置视角
    glViewport (0, 0, (GLsizei)w, (GLsizei)h);
    // 指定矩阵
    glMatrixMode (GL_PROJECTION);
    // 以相同矩阵取代现有矩阵
```

```cpp
    glLoadIdentity ();
    // 设置透视投影矩阵(perspective projection matrix)
    gluPerspective (60, (GLfloat)w / (GLfloat)h, 1.0, 100.0);
    glMatrixMode (GL_MODELVIEW);
}
int main (int argc, char **argv)
{
    // 启始 GLUT
    glutInit (&argc, argv);
    // 设置启始显示模式
    glutInitDisplayMode (GLUT_RGB | GLUT_DOUBLE);
    // 启始窗口大小
    glutInitWindowSize (200, 200);
    // 启始窗口位置
    glutInitWindowPosition (100, 100);
    // 建立窗口
    glutCreateWindow ("OpenCV Texture");
    // 显示回调函数
    glutDisplayFunc (display);
    // 程序让回调函数空闲
    glutIdleFunc (display);
    // 变形回调函数
    glutReshapeFunc (reshape);

    // 读取图文件
    Mat image;
    image = imread("c:/images/lena.jpg");

    // 载入图像到 OpenGL 纹理
    loadTexture(image, &texture);
    // GLUT 事件处理循环
    glutMainLoop ();

    // 释出纹理内存空间
    FreeTexture( texture );

    return 0;
}
```

OpenGL 的安装方法请参考附录，本程序是示范 OpenCV 与 OpenGL 整合。其执行结果如图 11-9 所示。

图 11-9

OpenGL 与摄像机

OpenGL 与摄像机应用的代码如下：

```cpp
#include <opencv2/core/core.hpp>
#include <opencv2/highgui/highgui.hpp>
#include "opencv2/imgproc/imgproc.hpp"
#include <iostream>
#include <glut.h>
using namespace cv;
using namespace std;
#define VIEWPORT_WIDTH          640
#define VIEWPORT_HEIGHT         480
#define KEY_ESCAPE              27
// 声明全局变量
VideoCapture cap(0);
GLint g_hWindow;
// 声明函数
GLvoid InitGL();
GLvoid OnDisplay();
GLvoid OnReshape(GLint w, GLint h);
GLvoid OnKeyPress (unsigned char key, GLint x, GLint y);
GLvoid OnIdle();
int main(int argc, char* argv[])
{
    // 建立 GLUT 窗口
    glutInit(&argc, argv);
    // 设置显示模式
    glutInitDisplayMode(GLUT_DOUBLE | GLUT_RGB | GLUT_DEPTH);
    // 设置窗口大小
    glutInitWindowSize(VIEWPORT_WIDTH, VIEWPORT_HEIGHT);
    // 建立窗口
    g_hWindow = glutCreateWindow("Video Texture");
    // 启动 OpenCV 摄像机
    if (!cap.isOpened())
    {
        cout << "无法启动摄像机" << endl;
        return -1;
    }
    // 启始 OpenGL
    InitGL();
    // Glut 显示循环
    glutMainLoop();
    return 0;
}
GLvoid InitGL()
{
    // 指定颜色,红,绿,蓝, alpha, 0 是起始值
    glClearColor (0.0, 0.0, 0.0, 0.0);
    // 显示的回调(callback)函数
    glutDisplayFunc(OnDisplay);
    // 窗口变形回调(callback)函数
    glutReshapeFunc(OnReshape);
    // 键盘回调(callback)函数
    glutKeyboardFunc(OnKeyPress);
```

```cpp
    // 全局等待(idle)回调(callback)函数
    glutIdleFunc(OnIdle);
}
GLvoid OnDisplay(void)
{
    glClear(GL_COLOR_BUFFER_BIT | GL_DEPTH_BUFFER_BIT);
    // 启动 OpenGL
    glEnable(GL_TEXTURE_2D);
    // 设置矩阵(Matrix)为 Projection
    glMatrixMode (GL_PROJECTION);
    glLoadIdentity();
    gluOrtho2D(0, VIEWPORT_WIDTH, VIEWPORT_HEIGHT, 0);
    // 转换矩阵为 Model View
    glMatrixMode(GL_MODELVIEW);
    glLoadIdentity();
    // 绘制纹理(textured)
    glBegin(GL_QUADS);
    glTexCoord2f(0.0f, 0.0f);
    glVertex2f(0.0f, 0.0f);
    glTexCoord2f(1.0f, 0.0f);
    glVertex2f(VIEWPORT_WIDTH, 0.0f);
    glTexCoord2f(1.0f, 1.0f);
    glVertex2f(VIEWPORT_WIDTH, VIEWPORT_HEIGHT);
    glTexCoord2f(0.0f, 1.0f);
    glVertex2f(0.0f, VIEWPORT_HEIGHT);
    glEnd();
    glFlush();
    glutSwapBuffers();
}
GLvoid OnReshape(GLint w, GLint h)
{
    // 设置窗口宽度与高度
    glViewport(0, 0, w, h);
}
GLvoid OnKeyPress(unsigned char key, int x, int y)
{
    switch (key) {
      case KEY_ESCAPE:
        glutDestroyWindow(g_hWindow);
        exit(0);
        break;
    }
    // 以现有窗口重新显示
    glutPostRedisplay();
}
GLvoid OnIdle()
{
    Mat frame;
    // 采集另一帧
    cap >> frame;
    // 转换成 RGB 格式
    cvtColor(frame, frame, CV_BGR2RGB);
    // 建立纹理(Texture)
    gluBuild2DMipmaps(GL_TEXTURE_2D, GL_RGB, frame.cols,
        frame.rows, GL_RGB, GL_UNSIGNED_BYTE, frame.data);
    // 更新显示(View port)
```

```
    glutPostRedisplay();
}
```

执行结果如图 11-10 所示。

图 11-10

看完本程序,如果读者对 OpenGL 程序设计有经验,应该可以设计扩增实境(Augmented Reality)的应用程序,因为本书是探讨 OpenCV 图像处理,这里不在此赘述扩增实境的范例应用。

附　　录

函数与章节对照表

为了方便快速查询使用过的函数，本书将 OpenCV 函数以字母顺序整理成如下的表 A-1。如果要查询函数属于哪个模块，可以在官方网站 http://docs.opencv.org/ 中单击每个模块的说明页面来查找。或者在搜索界面查询函数名称显示相关的查询结果。例如，cvtColor 查询的显示结果如图 A-1 所示。

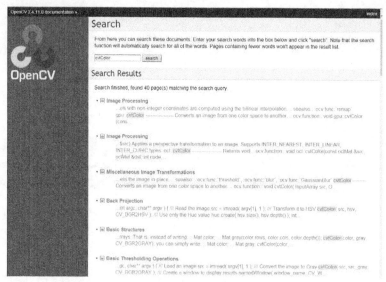

图 A-1

由查询结果可以知道，cvtColor 属于 Image Processing 模块，也就是 imgproc 模块。又查询到 Miscellaneous Image Transformations 表示此函数放在模块说明页的该组内。搜索完成之后，浏览代码即可知道要加入哪个 #include 文件以及项目属性程序库。

表 A-1

章　节	函数名称	函数功能
3.2	absdiff	计算两图差异的绝对值
2.3	add	计算两图像或颜色值相加
2.3	addWeighted	以权重将两图合并
4.20	approxPolyDP	指定精确度接近的多边曲线
4.1	bilateralFilter	左右对称过滤图像
11.1	bitwise_and	计算两个图像每个位的连接

续表

章 节	函数名称	函数功能
11.1	bitwise_or	计算两个图像每个位的分离
4.1	blur	归一化方格过滤来平滑图像
4.18	calcBackProject	计算直方图分布的反向投影
4.16	calcHist	计算一组阵列的直方图分布
7.4	calcOpticalFlowFarneback	用 Gunnar Farneback 的算法计算光流密度
4.10	Canny	用 Canny 算法寻找图像边缘
8.1	CascadeClassifier::detectMultiScale	在输入图像中检测不同大小的对象
8.1	CascadeClassifier::load	由文件载入级联式分类器
2.5	circle	绘制圆
10.1	clearBuffers	清除视网膜缓冲区
4.17	compareHist	直方图分布比较
4.8	convertScaleAbs	计算绝对值并将结果转换成 8 位
4.21	convexHull	寻找设置点的凸包
2.7	copyMakeBorder	在图像外围加上边框
3.1	copyTo	复制图像
6.6	cornerEigenValsAndVecs	于图像的区块（block）计算 eigen 值与向量，用来检测角
6.6	cornerHarris	哈瑞斯角点检测
6.7	cornerSubPix	精简角的地点
3.1	CreateTrackbar	建立窗口拉杆回调功能
9.1	CvSVM::train	训练 SVM
9.1	CvSVMParams	SVM 训练参数设置
2.2	cvtColor	图像颜色空间转换
2.7	dft	执行向前或反向的离散傅立叶转换
4.2	dilate	使用结构性元素膨胀图像
4.20	drawContours	绘制轮廓
6.8	drawKeypoints	绘制关键点
6.1	drawMatches	绘制两图之间发现匹配的关键点
2.5	ellipse	绘制椭圆
4.15	equalizeHist	直方图平等化
4.2	erode	使用结构性元腐蚀图像
6.1	FeatureDetector::detect	在图像中检测关键点
2.8	FileNode::begin	返回指向第一个节点元素的迭代器

续表

章　节	函数名称	函数功能
2.8	FileNode::end	返回指向最后节点元素的迭代器
2.8	FileStorage	FileStorage 类别构造函数
2.8	FileStorage& operator<<	将数据写入文件存储
2.8	FileStorage::open	打开文件
2.5	fillPoly	将多边形围绕区域填满
4.6	filter2D	用核心进行卷积处理
4.20	findContours	寻找二元图像的轮廓
11.2	flip	二维图像垂直、水平或两者都翻转
4.18	floodFill	用颜色填满连接的组件
3.2	GaussianBlur	用高斯过滤器平滑图像
9.1	get_support_vector_count	取得支持向量的数量
4.14	getAffineTransform	由三对点计算仿射转换
10.1	getMagno	视网膜移动色频的访问器
2.7	getOptimalDFTSize	由输入的向量大小取得最佳的 DFT 大小
10.1	getParvo	视网膜详尽色频的访问器
4.14	getRotationMatrix2D	计算 2D 旋转的仿射矩阵
2.6	getTextSize	计算文字符串的长与宽
6.5	goodFeaturesToTrack	决定图像的强转角
4.12	HoughCircles	用霍夫转换在二元图像中寻找圆
4.11	HoughLines	用霍夫转换在二元图像中寻找线
4.11	HoughLinesP	用机率的霍夫转换在二元图像中寻找线
2.1	imread	读取图文件
2.1	imshow	显示窗口
2.2	imwrite	存储图像
4.9	Laplacian	对图像进行拉普拉斯运算
2.5	line	绘制直线
2.7	log	对数运算
2.7	magnitude	计算二维向量的级数
11.2	Mat::clone	图像复制
4.19	matchTemplate	比较模板与图像重叠区域
4.1	medianBlur	用平均过滤器模糊图像
4.23	minAreaRec	寻找包围指定点最小区域的旋转矩形

续表

章 节	函数名称	函数功能
4.22	minEnclosingCircle	寻找包围指定点最小区域的圆
4.19	minMaxLoc	寻找图中最大与最小值
4.18	mixChannels	复制输入图像的指定色频到输出图像
4.24	moments	计算第三级多边形或点阵化的形状的时刻
4.3	morphologyEx	高级形态转换
2.1	namedWindow	建立要显示图档的窗口
2.7	normalize	矩阵值距或基准归一化处理
2.8	operator>>	由文件存储读取数据
6.4	perspectiveTransform	执行透视矩阵向量的转换
2.5	polylines	绘制多角形曲线
9.1	predict	预估输入样本后的反应
2.6	putText	在图像中绘制文字符串
4.4	pyrDown	向下取样（Upsamples）并模糊化
4.4	pyrUp	向上取样并模糊化
2.5	rectangle	绘制矩形
4.13	remap	对图像进行一般性的几何转换
2.3	resize	改变图像大小
10.1	Retina	视网膜构造函数
2.6	RNG	产生随机数
10.1	run	将视网膜功能应用到图像
4.8	Scharr	计算图像 X 或 Y 轴算子值
11.2	setMouseCallback	设置鼠标的回调函数
10.1	setup	调整视网膜参数设置
4.8	Sobel	计算第一、第二、第三或混合的图像算子
5.2	StereoBM::StereoBM	StereoBM 类的构造函数
4.5	threshold	阈值运算
7.1	VideoCapture cap(0)	指计算机连接的摄像机
7.2	VideoCapture::get	返回摄像机指定的特性
7.1	VideoCapture::read	读取摄影图像
7.2	VideoWrite::put	建立输出图像
7.2	VideoWriter::VideoWriter	建立输出图像
7.2	VideoWriter::write	写入摄影图像

续表

章 节	函 数 名 称	函 数 功 能
2.1	waitKey	等待按键
4.14	warpAffine	对图像执行仿射转换
10.1	write	将视网膜参数写入 xml 或 yml 格式的文件

安装 OpenGL

1. 下载

请在官方网站 https://www.opengl.org/resources/libraries/glut/ 下载，读者在下载时可能已有更新版本，请直接下载最新版本，如图 A-2 所示。

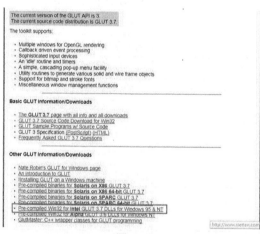

图 A-2

2. 解压缩文件

本书范例是解压缩到 C:\OpenGL 的路径下，读者也可以自行指定任何存放的文件夹。

3. 测试安装结果

OpenGL 相关函数说明，请参考说明文件 https://www.opengl.org/sdk/docs/man2/。

```
#include <glut.h>
// 画面显示
void RenderScene()
{
    // 清除窗口内容
    glClear(GL_COLOR_BUFFER_BIT);
    glFlush();
```

```
}
void main(void)
{
    // 显示模式
    glutInitDisplayMode(GLUT_SINGLE | GLUT_RGB);
    // 建立窗口
    glutCreateWindow("你好");
    // 显示回调函数
    glutDisplayFunc(RenderScene);
    // OpenGL 事件处理
    glutMainLoop();
}
```

执行结果如图 A-3 所示。

项目属性设置如图 A-4 所示。

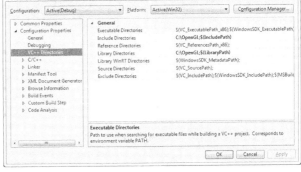

图 A-3　　　　　　　　　　　　　图 A-4

在按下 F5 键（start debugging）之前，将路径 C:\OpenGL 内的 glut32.dll 复制到项目的 debug 文件夹内，也就是要与执行文件位在相同路径。

读者也可以下载源代码来建立供 Visual Studio 使用的 Library，如图 A-5 所示。

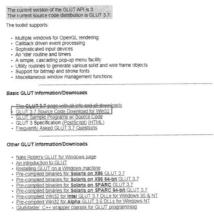

图 A-5

下载后请自行阅读 README.win 来建立供 Visual Studio 使用的 Library。

安装 Boost

  Boost 是网络上 C++ Library 的公开代码。虽然 HighGUI 模块只提供简单文件系统功能，但是笔者发现 Boost 在管理文件目录相关的功能还不错，所以在本附录中包含了其安装说明。

1. 下载

  请在官方网站 http://www.boost.org/ 下载。读者下载时可能已有更新版本，请直接下载最新版本，如图 A-6 所示。

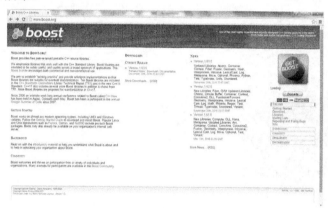

图 A-6

2. 解压缩下载文件

  本书范例是解压缩到 C:\路径下，读者也可指定任意存放的文件夹，如图 A-7 所示。

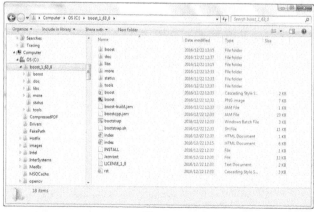

图 A-7

3. 打开命令提示字符

单击 Windows 左下角的"开始",找到 Visual Studio 2013,再单击 Visual Studio Tools,如图 A-8 所示。

图 A-8

然后再单击"VS2013 开发人员命令提示字符",如图 A-9 所示。

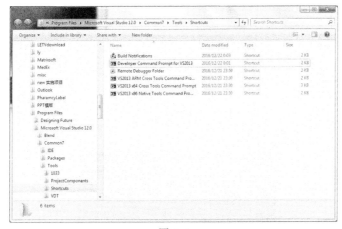

图 A-9

修改更换到 C:\boost_1_57_0 或是自行更换解压缩后的文件夹。

4. 执行 bootstrap

选择 bootstrap,如图 A-10 所示。

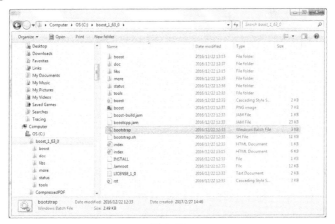

图 A-10

5. 再执行 b2

再选择 b2，如图 A-11 所示。

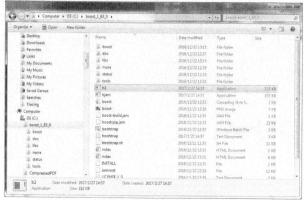

图 A-11　选择 b2

此步骤要花一些时间，执行完成后的结果如图 A-12 所示。

图 A-12

在 C:\boost_1_57_0 的文件夹内产生一个名称为 stage 文件夹，其内容就是供 Visual Studio 使用的程序库，如图 A-13 所示。

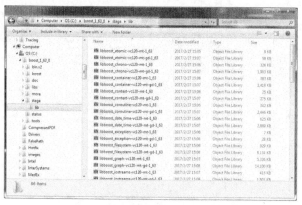

图 A-13

6. 测试安装结果

还是使用 Visual C++ Win32 Console Application，本书所有的范例都是这样。

```cpp
#include <boost/filesystem.hpp>
#include <iostream>
using namespace std;
using namespace boost::filesystem;
int main() {
    string PATH("C:/images");
    for(recursive_directory_iterator i = recursive_directory_iterator(PATH),
                                    end_iter; i != end_iter; i++)
    {
        if (i.level() > 0)
        {
            for (int k = 1; k <= i.level(); k++)
                printf(" ");
        }
        cout << i.level() << " " << (i->path()).filename().string() << endl;
    }
    getchar();
    return 0;
}
```

项目属性设置如图 A-14 所示。

图 A-14

执行结果，如果看到文件夹内容就成功了。该程序是指定到本书使用到的图文件目录，读者可自行更改程序到任何文件夹下。执行结果如图 A-15 所示。

图 A-15

欢迎来到异步社区！

异步社区的来历

异步社区（www.epubit.com.cn）是人民邮电出版社旗下 IT 专业图书旗舰社区，于 2015 年 8 月上线运营。

异步社区依托于人民邮电出版社 20 余年的 IT 专业优质出版资源和编辑策划团队，打造传统出版与电子出版和自出版结合、纸质书与电子书结合、传统印刷与 POD 按需印刷结合的出版平台，提供最新技术资讯，为作者和读者打造交流互动的平台。

社区里都有什么？

购买图书

我们出版的图书涵盖主流 IT 技术，在编程语言、Web 技术、数据科学等领域有众多经典畅销图书。社区现已上线图书 1000 余种，电子书 400 多种，部分新书实现纸书、电子书同步出版。我们还会定期发布新书书讯。

下载资源

社区内提供随书附赠的资源，如书中的案例或程序源代码。

另外，社区还提供了大量的免费电子书，只要注册成为社区用户就可以免费下载。

与作译者互动

很多图书的作译者已经入驻社区，您可以关注他们，咨询技术问题；可以阅读不断更新的技术文章，听作译者和编辑畅聊好书背后有趣的故事；还可以参与社区的作者访谈栏目，向您关注的作者提出采访题目。

灵活优惠的购书

您可以方便地下单购买纸质图书或电子图书，纸质图书直接从人民邮电出版社书库发货，电子书提供多种阅读格式。

对于重磅新书，社区提供预售和新书首发服务，用户可以第一时间买到心仪的新书。

用户账户中的积分可以用于购书优惠。100 积分 =1元，购买图书时，在 使用积分 里填入可使用的积分数值，即可扣减相应金额。

特 别 优 惠

购买本书的读者专享异步社区购书优惠券。

使用方法：注册成为社区用户，在下单购书时输入 S4XC5 使用优惠码 ，然后点击"使用优惠码"，即可在原折扣基础上享受全单9折优惠。（订单满39元即可使用，本优惠券只可使用一次）

纸电图书组合购买

社区独家提供纸质图书和电子书组合购买方式，价格优惠，一次购买，多种阅读选择。

社区里还可以做什么？

提交勘误

您可以在图书页面下方提交勘误，每条勘误被确认后可以获得 100 积分。热心勘误的读者还有机会参与书稿的审校和翻译工作。

写作

社区提供基于 Markdown 的写作环境，喜欢写作的您可以在此一试身手，在社区里分享您的技术心得和读书体会，更可以体验自出版的乐趣，轻松实现出版的梦想。

如果成为社区认证作译者，还可以享受异步社区提供的作者专享特色服务。

会议活动早知道

您可以掌握 IT 圈的技术会议资讯，更有机会免费获赠大会门票。

加入异步

扫描任意二维码都能找到我们：

异步社区　　微信服务号　　微信订阅号　　官方微博　　QQ 群：436746675

社区网址：www.epubit.com.cn

投稿 & 咨询：contact@epubit.com.cn